Gene Probes

METHODS IN MOLECULAR BIOLOGY™

John M. Walker, Series Editor

190. **High Throughput Screening:** *Methods and Protocols,* edited by *William P. Janzen, 2002*
189. **GTPase Protocols:** *The RAS Superfamily,* edited by *Edward J. Manser and Thomas Leung, 2002*
188. **Epithelial Cell Culture Protocols,** edited by *Clare Wise, 2002*
187. **PCR Mutation Detection Protocols,** edited by *Bimal D. M. Theophilus and Ralph Rapley, 2002*
186. **Oxidative Stress and Antioxidant Protocols,** edited by *Donald Armstrong, 2002*
185. **Embryonic Stem Cells:** *Methods and Protocols,* edited by *Kursad Turksen, 2002*
184. **Biostatistical Methods,** edited by *Stephen W. Looney, 2002*
183. **Green Fluorescent Protein:** *Applications and Protocols,* edited by *Barry W. Hicks, 2002*
182. **In Vitro Mutagenesis Protocols, 2nd ed.**, edited by *Jeff Braman, 2002*
181. **Genomic Imprinting:** *Methods and Protocols,* edited by *Andrew Ward, 2002*
180. **Transgenesis Techniques, 2nd ed.:** *Principles and Protocols,* edited by *Alan R. Clarke, 2002*
179. **Gene Probes:** *Principles and Protocols,* edited by *Marilena Aquino de Muro and Ralph Rapley, 2002*
178.`**Antibody Phage Display:** *Methods and Protocols,* edited by *Philippa M. O'Brien and Robert Aitken, 2001*
177. **Two-Hybrid Systems:** *Methods and Protocols,* edited by *Paul N. MacDonald, 2001*
176. **Steroid Receptor Methods:** *Protocols and Assays,* edited by *Benjamin A. Lieberman, 2001*
175. **Genomics Protocols**, edited by *Michael P. Starkey and Ramnath Elaswarapu, 2001*
174. **Epstein-Barr Virus Protocols,** edited by *Joanna B. Wilson and Gerhard H. W. May, 2001*
173. **Calcium-Binding Protein Protocols, Volume 2:** *Methods and Techniques,* edited by *Hans J. Vogel, 2001*
172. **Calcium-Binding Protein Protocols, Volume 1:** *Reviews and Case Histories,* edited by *Hans J. Vogel, 2001*
171. **Proteoglycan Protocols,** edited by *Renato V. Iozzo, 2001*
170. **DNA Arrays:** *Methods and Protocols,* edited by *Jang B. Rampal, 2001*
169. **Neurotrophin Protocols,** edited by *Robert A. Rush, 2001*
168. **Protein Structure, Stability, and Folding,** edited by *Kenneth P. Murphy, 2001*
167. **DNA Sequencing Protocols,** *Second Edition,* edited by *Colin A. Graham and Alison J. M. Hill, 2001*
166. **Immunotoxin Methods and Protocols**, edited by *Walter A. Hall, 2001*
165. **SV40 Protocols**, edited by *Leda Raptis, 2001*
164. **Kinesin Protocols,** edited by *Isabelle Vernos, 2001*
163. **Capillary Electrophoresis of Nucleic Acids, Volume 2:** *Practical Applications of Capillary Electrophoresis,* edited by *Keith R. Mitchelson and Jing Cheng, 2001*
162. **Capillary Electrophoresis of Nucleic Acids, Volume 1:** *Introduction to the Capillary Electrophoresis of Nucleic Acids,* edited by *Keith R. Mitchelson and Jing Cheng, 2001*
161. **Cytoskeleton Methods and Protocols**, edited by *Ray H. Gavin, 2001*
160. **Nuclease Methods and Protocols,** edited by *Catherine H. Schein, 2001*
159. **Amino Acid Analysis Protocols,** edited by *Catherine Cooper, Nicole Packer, and Keith Williams, 2001*
158. **Gene Knockout Protocols,** edited by *Martin J. Tymms and Ismail Kola, 2001*
157. **Mycotoxin Protocols,** edited by *Mary W. Trucksess and Albert E. Pohland, 2001*
156. **Antigen Processing and Presentation Protocols,** edited by *Joyce C. Solheim, 2001*
155. **Adipose Tissue Protocols,** edited by *Gérard Ailhaud, 2000*
154. **Connexin Methods and Protocols,** edited by *Roberto Bruzzone and Christian Giaume, 2001*
153. **Neuropeptide Y Protocols**, edited by *Ambikaipakan Balasubramaniam, 2000*
152. **DNA Repair Protocols:** *Prokaryotic Systems*, edited by *Patrick Vaughan, 2000*
151. **Matrix Metalloproteinase Protocols,** edited by *Ian M. Clark, 2001*
150. **Complement Methods and Protocols,** edited by *B. Paul Morgan, 2000*
149. **The ELISA Guidebook**, edited by *John R. Crowther, 2000*
148. **DNA–Protein Interactions:** *Principles and Protocols* **(2nd ed.)**, edited by *Tom Moss, 2001*
147. **Affinity Chromatography:** *Methods and Protocols*, edited by *Pascal Bailon, George K. Ehrlich, Wen-Jian Fung, and Wolfgang Berthold, 2000*
146. **Mass Spectrometry of Proteins and Peptides,** edited by *John R. Chapman, 2000*
145. **Bacterial Toxins:** ***Methods and Protocols***, edited by *Otto Holst, 2000*
144. **Calpain Methods and Protocols**, edited by *John S. Elce, 2000*
143. **Protein Structure Prediction:** *Methods and Protocols*, edited by *David Webster, 2000*
142. **Transforming Growth Factor-Beta Protocols**, edited by *Philip H. Howe, 2000*
141. **Plant Hormone Protocols**, edited by *Gregory A. Tucker and Jeremy A. Roberts, 2000*
140. **Chaperonin Protocols**, edited by *Christine Schneider, 2000*
139. **Extracellular Matrix Protocols**, edited by *Charles Streuli and Michael Grant, 2000*
138. **Chemokine Protocols**, edited by *Amanda E. I. Proudfoot, Timothy N. C. Wells, and Christine Power, 2000*
137. **Developmental Biology Protocols, Volume III**, edited by *Rocky S. Tuan and Cecilia W. Lo, 2000*
136. **Developmental Biology Protocols, Volume II,** edited by *Rocky S. Tuan and Cecilia W. Lo, 2000*
135. **Developmental Biology Protocols, Volume I,** edited by *Rocky S. Tuan and Cecilia W. Lo, 2000*
134. **T Cell Protocols:** *Development and Activation*, edited by *Kelly P. Kearse, 2000*
133. **Gene Targeting Protocols,** edited by *Eric B. Kmiec, 2000*
132. **Bioinformatics Methods and Protocols,** edited by *Stephen Misener and Stephen A. Krawetz, 2000*
131. **Flavoprotein Protocols,** edited by *S. K. Chapman and G. A. Reid, 1999*

METHODS IN MOLECULAR BIOLOGY™

Gene Probes

Principles and Protocols

Edited by

Marilena Aquino de Muro

CABI Bioscience UK Centre, Surrey, UK

and

Ralph Rapley

University of Hertfordshire, Hertfordshire, UK

Humana Press Totowa, New Jersey

Softcover reprint of the hardcover 1st edition 2002
999 Riverview Drive, Suite 208
Totowa, New Jersey 07512

This publication is printed on acid-free paper. ∞
ANSI Z39.48-1984 (American Standards Institute) Permanence of Paper for Printed Library Materials.

Cover Illustration:

Cover design by Patricia F. Cleary.

Production Editor: Mark J. Breaugh.

For additional copies, pricing for bulk purchases, and/or information about other Humana titles, contact Humana at the above address or at any of the following numbers: Tel: 973-256-1699; Fax: 973-256-8341; E-mail: humana@humanapr.com, or visit our Website at www.humanapress.com

10 9 8 7 6 5 4 3 2 1

Library of Congress Cataloging in Publication Data

Gene probes : principles and protocols / edited by Marilena Aquino de Muro and Ralph Rapley.
p. cm. -- (Methods in Molecular Biology ; 179)
Includes bibliographical references and index.
ISBN 978-1-61737-236-0 e-ISBN 978-1-59259-238-8
1. Nucleic acid probes--Laboratory manuals. I. Aquino deMuro, Marilena. II. Rapley, Ralph. III. Series.

QP624.5.D73 G465 2001
572.8--dc21

2001024303

Preface

The use of nucleic acid or gene probes as cloning or diagnostic tools has proven to be a powerful technique for mapping, isolation, and manipulation of genes. Their applications range from the identification of genes responsible for inherited human disorders to the rapid detection and identification of bacteria or other microorganisms in microbial communities. For example, oligonucleotide hybridization probes complementing either small ribosomal subunits, large ribosomal subunits, or internal transcribed spacer regions have now been developed for a wide variety of microorganisms. The technique of *in situ* hybridization of labeled DNA has allowed genes to be mapped to single chromosomes and in many cases to a single chromosome band, promoting significant advance in human genome mapping.

Gene Probes: Principles and Protocols presents the principles for gene probe design, labeling, detection, target format, and hybridization conditions together with detailed protocols, accompanied by practical hints and tips from experts. It will be a valuable resource for all those bench scientists—including biochemists, molecular biologists, and microbiologists—both in academia and in industry who are engaged in the search for specific genes in the human chromosome or in detection of microorganisms and their toxins, both in the environment and in food samples, to mention only two examples.

Marilena Aquino de Muro
Ralph Rapley

Contents

Contributors

SALLY A. ANDERSON • *School of Biological Sciences, University of Auckland, Auckland, New Zealand*
MARILENA AQUINO DE MURO • *CABI Bioscience UK Centre, Surrey, UK*
TERESA BISUCCI • *Department of Human Biology and Movement Science, RMIT University, Bundoora, Victoria, Australia*
LINDA L. BLACKALL • *Advanced Wastewater Management Centre, Department of Microbiology and Parasitology, The University of Queensland, Queensland, Australia*
CLAUDIO BONGHI • *Department of Environmental Agronomy and Crop Science, University of Padova, Padova, Italy*
ROSS D. BROWN • *Institute of Hematology, Royal Prince Alfred Hospital, Camperdown, Australia*
JONGSIK CHUN • *University of Maryland and Biotechnology Institute, Center of Marine Biotechnology, Baltimore, MD*
RITA R. COLWELL • *University of Maryland and Biotechnology Institute, Center of Marine Biotechnology, Baltimore, MD*
CARLOS CORDON-CARDO • *Department of Pathology, Memorial Sloan-Kettering Cancer Center, New York, NY*
IAN A. DARBY • *Department of Human Biology and Movement Science, RMIT University, Bundoora, Victoria, Australia*
MICHAEL R. DAVEY • *Plant Genetic Manipulation Group, Department of Life Science, University of Nottingham, Nottingham, UK*
HUBERT ENDTZ • *Erasmus University Medical Center Rotterdam, Department of Medical Microbiology and Infectious Diseases, Rotterdam, The Netherlands*
LEE C. GARRATT • *Plant Genetic Manipulation Group, Department of Life Science, University of Nottingham, Nottingham, UK*
TIM D. HEWITSON • *Department of Nephrology, Royal Melbourne Hospital, Parkville, Australia*
ELENA HILARIO • *HortResearch, The Horticulture and Food Research Institute of New Zealand Ltd., Auckland, New Zealand*
P. JOY HO • *Institute of Hematology, Royal Prince Alfred Hospital, Camperdown, Australia*

PHILIP HUGENHOLTZ • *Advanced Computational Modeling Centre, Department of Mathematics, The University of Queensland, Queensland, Australia*
KEITH A. LAMPEL • *Center for Food Safety and Applied Nutrition, Food and Drug Administration, Washington, DC*
ALICE C. LAYTON • *Center for Environmental Biotechnology and Department of Microbiology, The University of Tennessee, Knoxville, TN*
LAURA G. LEFF • *Department of Biological Sciences, Kent State University, Kent, OH*
GILLIAN D. LEWIS • *School of Biological Sciences, University of Auckland, Auckland, New Zealand*
LUIS LUIS • *Laboratorio de Bioquimica y Biologia Molecular de Parasitos, Instituto de Biologia Experimental, Universidade Central de Venezuela, Caracas, Venezuela*
CLARA MARTÍNEZ • *Laboratorio de Bioquimica y Biologia Molecular de Parasitos, Instituto de Biologia Experimental, Universidade Central de Venezuela, Caracas, Venezuela*
SARAH MATHEWS • *Centre for Molecular Biotechnology and CRC for Diagnostic Technologies, Queensland University of Technology, Brisbane, Queensland, Australia*
MATTHEW S. MCCABE • *Plant Genetic Manipulation Group, Department of Life Science, University of Nottingham, Nottingham, UK*
CHRISTOPHER J. MCNAMARA • *Division of Engineering and Applied Sciences, Harvard University, Cambridge, MA*
ALEXIS MENDOZA-LEÓN • *Laboratorio de Bioquimica y Biologia Molecular de Parasitos, Instituto de Biologia Experimental, Universidade Central de Venezuela, Caracas, Venezuela*
J. COLIN MURRELL • *Department of Biological Sciences, University of Warwick, Coventry, UK*
GERARD J. NUOVO • *Department of Pathology, The Ohio State University Medical Center, Columbus, OH*
PALMER A. ORLANDI • *Center for Food Safety and Applied Nutrition, Food and Drug Administration, Washington, DC*
IRENE ORLOW • *Department of Epidemiology and Biostatistics, Memorial Sloan-Kettering Cancer Center, New York, NY*
J. BRIAN POWER • *Plant Genetic Manipulation Group, Department of Life Science, University of Nottingham, Nottingham, UK*
STEFAN RADAJEWSKI • *Department of Biological Sciences, University of Warwick, Coventry, UK*

Ralph Rapley • *Department of Biosciences, University of Hertfordshire, Hertfordshire, UK*
Alex Reid • *CABI Bioscience UK Centre, Surrey, UK*
Irma N.G. Rivera • *University of Maryland and Biotechnology Institute, Center of Marine Biotechnology, Baltimore, MD*
Benedetto Ruperti • *Department of Environmental Agronomy and Crop Science, University of Padova, Padova, Italy*
John Sanseverino • *Center for Environmental Biotechnology and Department of Microbiology, The University of Tennessee, Knoxville, TN*
Gary S. Sayler • *Center for Environmental Biotechnology and Department of Microbiology, The University of Tennessee, Knoxville, TN*
Peter Timms • *Centre for Molecular Biotechnology and CRC for Diagnostic Technologies, Queensland University of Technology, Brisbane, Queensland, Australia*
Pietro Tonutti • *Department of Environmental Agronomy and Crop Science, University of Padova, Padova, Italy*
Susan J. Turner • *School of Biological Sciences, University of Auckland, Auckland, New Zealand*
Gene W. Tyson • *Advanced Wastewater Management Centre, Department of Microbiology and Parasitology, The University of Queensland, Queensland, Australia*
Alex van Belkum • *Erasmus University Medical Center Rotterdam, Department of Medical Microbiology and Infectious Diseases, Rotterdam, The Netherlands*
Nicole van den Braak • *Erasmus University Medical Center Rotterdam, Department of Medical Microbiology and Infectious Diseases, Rotterdam, The Netherlands*
Leen-Jan van Doorn • *Delfs Diagnostics Laboratories, Delfs, The Netherlands*
Debra K. Winters • *Department of Food Science and Biotechnology Center, University of Arkansas, Fayetteville, AR*

1

Target Format and Hybridization Conditions

Alex Reid

1. Southern Blotting

1.1. Introduction

The isolation of specific regions within the genome of an organism is now normally accomplished by polymerase chain reaction (PCR) amplification using primers specific for the region in question. However, there are occasions where this is not possible (loss of primer sites resulting in no amplification or if there are no primers available). In these cases the detection of the sequence of interest can be achieved by hybridization of a labeled probe to restricted genomic DNA immobilized on a membrane by Southern blotting ***(1)***. Genomic DNA is first digested by one or more restriction enzymes and the fragments generated separated by gel electrophoresis. The amount of DNA to be applied to the gel varies from application to application. In general 10 μg of human genomic DNA is needed for the detection of a single copy gene when using radioactively labeled probes and an overnight exposure to X-ray film. This figure can be reduced if the target is either a repetitive element (e.g., ribosomal DNA) or, if plasmid DNA or PCR products are run on the gel. Once the fragments are separated on the gel the DNA is then denatured *in situ* and transferred by capillary transfer to either a nitrocellulose or nylon membrane. The DNA fragments are then bound to the membrane, which can then be used in a hybridization reaction.

1.2. Materials

1.2.1. Specific Materials

1. 3MM filter paper (Whatman), paper towels, glass or plastic tray and support (a gel casting tray turned upside down), cling film.

From: *Methods in Molecular Biology, vol. 179: Gene Probes: Principles and Protocols*
Edited by: M. Aquino de Muro and R. Rapley © Humana Press Inc., Totowa, NJ

2. Hybond N membrane (Amersham Pharmacia Biotech).
3. Denaturing solution: 1.5 *M* NaCl, 0.5 *M* NaOH.
4. Neutralizing solution: 1.5 *M* NaCl, 0.5 *M* Tris-HCl, pH 7.2, 0.001 *M* EDTA.
5. 20× Saline sodium citrate (SSC): 3 *M* NaCl, 0.3 *M* Na_3 citrate.
6. UV transilluminator.
7. 100× Denhardt's solution: 10 g of bovine serum albumin (BSA) fraction V, 10 g of Ficoll 400, 10 g of polyvinylpyrrolidone (PVP) in 500 mL of distilled water. Store at –20°C in 10 mL aliquots.
8. 10% (w/v) Sodium dodecyl sulfate (SDS).
9. 10 mg/mL sheared herring testis DNA. Store at –20°C.

1.2.2. Optional Materials

1. 0.25 *M* HCl.
2. 0.4 *M* NaOH and a solution of 0.1× SSC, 0.1% (w/v) SDS, and 0.2 *M* Tris-HCl, pH 7.5.
3. Oven set at 80°C.
4. Vacuum blotting system (VacuGene system from Amersham Pharmacia Biotech).

1.3. Method

1.3.1. Preparation of the Gel for Transfer

1. Electrophorese samples in an agarose gel (*see* **Note 1**) and transfer the gel to a glass or plastic tray slightly larger than the gel.
2. If the fragments for analysis are large (>10 kb) the efficiency of transfer can be increased by depurinating the DNA. Add 0.25 *M* HCl to the tray containing the gel until the gel is just covered (*see* **Note 2**). Place on a rocking platform or orbital shaker and agitate gently for 15–25 min at room temperature.
3. Remove the 0.25 *M* HCl, rinse the gel with distilled water and cover with denaturing solution. Return the tray to the rocker and shake for 30 min at room temperature (or 15 min after the dye has returned to blue).
4. Remove the denaturing solution, rinse the gel with distilled water and cover with neutralizing solution. Shake for 15 min at room temperature.
5. Repeat with fresh neutralizing solution.
6. Set up the capillary blot.

1.3.2. Setting Up the Capillary Blot

1. Half fill a glass or plastic tray of a suitable size with 20× SSC (*see* **Note 3**). Place a support in the tray (the upturned casting tray in which the gel was cast). Cover the support with a wick made from three sheets of 3MM paper. Allow the 20× SSC to wet the wick and ensure there are no air bubbles trapped between the sheets and the support (*see* **Note 4**).
2. Carefully place the treated gel on the wick ensuring there are no air bubbles

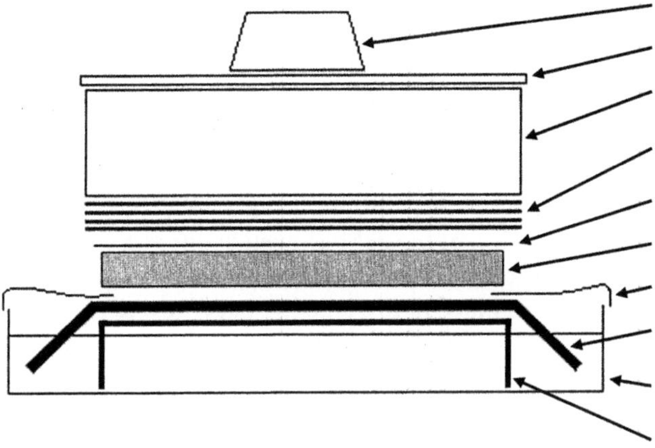

Figure 1. Schematic of a capillary Southern blot.

between the wick and the gel. Surround the gel with cling film (**Fig. 1**) to ensure that transfer occurs through the gel and not around the sides.

3. Place the membrane onto the gel (*see* **Note 5**).
4. Wet two sheets of 3MM paper cut slightly larger than the membrane with 2× SSC and place these (one at a time) on top of the membrane ensuring there are no bubbles.
5. Place a dry sheet of 3MM paper on top of the wet ones. Repeat with another sheet of 3MM paper.
6. Place a stack of paper towels 5–10 cm high on top of the 3MM paper and cover with a glass plate. Put a 500-g weight on top of the glass plate.
7. Allow transfer to proceed for several hours (preferably overnight).
8. After blotting carefully dismantle the stack of paper towels, 3MM sheets, etc. to expose the membrane. Before removing the membrane mark the edges of the gel with a pencil (if desired the wells can also be marked).
9. Remove the membrane and rinse carefully in 2× SSC to remove any adhering pieces of agarose.
10. Air-dry the membrane on a sheet of 3MM paper.
11. Fix the DNA to the membrane either by baking at 80°C for 2 h or by wrapping the membrane in cling film and placing DNA side down on a UV transilluminator for 2–5 min (*see* **Note 6**).

1.4. Optional Methods for DNA Transfer

1.4.1. Bidirectional Transfer to Two Membranes

If required the DNA in a gel can simultaneously be transferred to two membranes using the method of Smith and Summer *(2)*. This method is of benefit if many probes need to be hybridized to the DNA in a short space of time.

1. Prepare the gel for transfer as outlined in **Subheading 1.3**.
2. After the final neutralization step cover the gel in 10× SSC and shake for 30 min.
3. Wet two sheets of 3MM paper in 2× SSC. Place one on a flat clean surface.
4. Place a nylon membrane on top of the sheet of 3MM paper, ensuring there are no air bubbles between the 3MM and the membrane.
5. Carefully place the gel on top of the membrane. Do not move the gel once it is in contact with the membrane as transfer will start immediately.
6. Place the second membrane on top of the gel followed by the second wet sheet of 3MM paper, again making sure there are no air bubbles.
7. Pick up the "gel sandwich" and place onto a stack of paper towels. Cover the top with a similar stack of towels. Place a sheet of glass on the top and weigh down as before.
8. Allow transfer to proceed as before.

1.4.2. Vacuum Blotting

There are a number of alternative methods for DNA transfer from agarose gels to membranes. One of the best of these in terms of simplicity and speed is vacuum blotting. Here the DNA is literally sucked out of the gel onto the membrane and the entire process can be carried out between 20 and 60 min. Using a vacuum blotting system several gels can be blotted in a single day. The system used in our laboratory is the VacuGene XL available from Amersham Pharmacia Biotech.

1. Set up the vacuum blot apparatus ensuring there is a liquid trap between the pump and the blotter.
2. Prewet the support screen with distilled water and place shiny side up in the blotter.
3. Place a plastic mask on the support screen with a precut hole slightly smaller than the gel to be blotted.
4. Position the membrane over the hole in the mask, ensuring there are no air bubbles between the membrane and the support screen.
5. Carefully place the gel on top of the membrane, ensuring that there are no air bubbles between the gel and the membrane and that the edges of the gel protrude over the hole in the plastic mask.
6. Clamp the top of the blotting apparatus to the lower part containing the gel.
7. Switch on the vacuum pump and pour 0.25 *M* HCl into the apparatus so that it covers the gel. Stabilize the vacuum at 50 mbar and leave for 4 min.
8. Remove the 0.25 *M* HCl by tilting the apparatus and sucking off the solution. This can be achieved by having a "T" connector between the liquid trap and the blotter, which can be opened and closed by means of a clip. Residual solutions can be removed from the gel surface by wiping with a gloved finger or a disposable pipet.
9. Pour in the denaturing solution until it covers the gel and leave for 3 min. Remove as before.

10. Cover the gel with neutralizing solution (1.0 *M* Tris-HCl, pH 5.0, 1.5 *M* NaCl, 0.001 *M* EDTA), leave for 3 min, and remove.
11. Cover the gel with 20× SSC and leave for 20–60 min (*see* **Note 7**). Make sure the gel remains immersed during transfer.
12. Remove the 20× SSC and with the vacuum still applied peel the gel off the membrane. Switch off the vacuum and remove the filter. Treat as before.

1.5. Hybridization Conditions

There are many different hybridization solutions in the literature. The one detailed here is simple to make and gives low background with the Hybond N nylon membranes. The hybridization can be carried out in either heat sealed plastic bags that can withstand the necessary temperatures or in plastic boxes with sealable lids.

1. Make up a prehybidization solution that contains final concentrations of 5× SSC, 5× Denhardt's solution, and 0.5% SDS. Allow 125 μL of solution per cm^2 of membrane. Place the prehybridization solution in a 50-mL tube and place in a 65°C water bath.
2. Boil enough herring testis DNA to give a final concentration of 100 μg/mL for 5 min and snap cool on ice. Add to the prehybridization solution.
3. Prewet the membrane to be hybridized in 5× SSC and place in an opened out plastic bag (*see* **Note 8**). Close the bag over the filter and heat seal around the edges as close to the gel as possible leaving the top open.
4. Pour in the prehybridization solution, squeeze out as much air as possible, and seal the top of the bag with a heat sealer.
5. Place the bag between two sheets of glass and place in a shaking 65°C water bath. Incubate for at least 30 min.
6. Denature the labeled probe by boiling for 5 min and snap cooling on ice. Cut one corner off the hybridization bag and pipet the probe into the prehybridization solution. Reseal the bag and incubate at 65°C in a shaking water bath overnight.
7. Prepare wash solutions (1–5 mL/cm^2 membrane) and preheat to 65°C.
8. At the end of the hybridization carefully cut one corner off the bag and pour the hybridization solution into suitable container for disposal. Open the bag and remove the membrane and place in a sealable plastic box.
9. Wash the membrane by incubating at 65°C in a shaking water bath in the following solutions: 2× SSC, 0.1% SDS for 5 min (repeat), 1× SSC, 0.1% SDS for 15 min, 0.1× SSC, 0.1% SDS for 10 min (repeat) (*see* **Note 9**).
10. Remove the membrane from the last wash solution and drain the excess liquid off. Wrap in cling film and expose to X-ray film (*see* **Note 10**). If the membrane is to be reprobed it must be kept moist.

1.6. Probe Removal from Nylon Membranes

If the membrane needs to be hybridized with more than one probe the old probe can be removed from nylon membranes (providing they have not dried out) using the following procedure.

1. Place the membrane in a sealable plastic box. Cover the membrane with 0.4 *M* NaOH and incubate at 45°C for 30 mins in a shaking water bath.
2. Pour off the NaOH solution and cover the membrane with 0.1× SSC, 0.1% SDS, 0.2 *M* Tris-HCl, pH 7.5. Incubate at 45°C for 15 min in a shaking water bath.
3. Wrap the membrane in cling film and expose to X-ray film to ensure the old probe has been removed.
4. Filters can be stored wrapped in cling film at –20°C indefinitely.

1.7. Notes

1. For genomic DNA transfer use agarose (type I; low EEO from Sigma) as this allows good transfer of the DNA out of the gel and is fairly cheap.
2. When the samples are loaded on the gel use a loading buffer containing 0.25% (w/v) bromophenol blue. The depurination step can be monitored by the change in color of the dye from blue to yellow. Once the dye has changed color leave the gel for an additional 10 min.
3. An alternative transfer buffer is 20× SSPE: 3.6 *M* NaCl, 0.2 *M* sodium phosphate, 0.02 *M* EDTA, pH 7.7. However, 20× SSC is cheaper.
4. Air bubbles are easily removed by rolling a disposable pipet gently over the surface. This method can be used at all stages of set up.
5. Small DNA fragments will start to transfer to the membrane immediately on contact. Therefore, do not move the membrane once it establishes contact with the gel. Nylon membranes do not require pre wetting before application to the gel. If a nitrocellulose membrane is to be used float the membrane on the surface of a tray filled with distilled water until it is completely wet. Carefully immerse the membrane and leave for 5 min.
6. The optimum exposure time varies between transilluminators and can also change with the age of the UV bulbs. To calibrate the transilluminator run a gel containing six lanes with 50 pg of λ DNA digested with *Hin*dIII in each lane. Blot the DNA onto a membrane and cut the filter into six strips. Expose each strip for varying lengths of time ranging from 30 s to 10 min. Hybridize these blots to λ DNA and expose to X-ray film. The optimum exposure time can be determined by the strip which gives the strongest signal.
7. Transfer times vary depending on the thickness and concentration of the gel, the size of the fragments to be transferred, and the level of vacuum applied.
8. Several membranes can be hybridized in the same bag with little loss of signal.
9. If a radioactive probe is used the progress of the washes can be monitored using a hand held counter. The membrane is ready for autoradiography when the counts fall to near background on areas of the membrane containing no DNA. If in doubt stop the washes early and expose to X-ray film. It is always possible to wash the membrane further if the signal is too strong.
10. Exposure time vary depending on the amount of DNA run on the gel, the specific activity and nature of the probe. A probe hybridizing to a single copy sequence will require longer exposure time than one for a repetitive element. The optimum exposure time will need to be determined for each experiment. In general, an

overnight exposure should suffice for most applications. Exposure times can be shortened by preflashing the X-ray film. To do this mount a flash gun on a support about 50 cm above the bench in the darkroom. Cover the lens of the flash gun with several layers of paper to reduce the amount of light emitted. Take a piece of X-ray film and place below the flash gun. Cover four fifths of the film with a sheet of card and fire the flash gun. Move the sheet of card so that three fifths of the film is exposed and fire the flash gun again. Repeat until the entire film is exposed. Develop the film to determine the optimum flash time for the film/flash gun combination. The required exposure does not alter the background of the film whereas the next exposure does. Flashing the X-ray film in this way presensitizes the film, thus reducing exposure times.

2. Slot/Dot Blots

2.1. Introduction

If large numbers of samples need to by hybridized to a probe that yields a positive/negative result (such as species specific clones), this can be achieved by dot or slot blotting. Using this technique DNA is applied directly to a membrane and therefore no gel electrophoresis is required. Commercial manifolds are available which can be attached to a vacuum source where the DNA is applied to wells from which it is sucked onto the membrane in an ordered array. Alternatively, the DNA can be pipetted directly onto the membrane using a micropipet.

2.2. Materials

2.2.1. Specific Materials

1. 3MM filter paper (Whatman).
2. Hybond N membrane (Amersham Pharmacia Biotech).
3. Denaturing solution: 1.5 *M* NaCl, 0.5 *M* NaOH.
4. Neutralizing solution: 1.5 *M* NaCl, 0.5 *M* Tris-HCl, pH 7.2, 0.001 *M* EDTA.
5. 20× SSC: 3 *M* NaCl, 0.3 *M* Na_3 citrate.
6. UV transilluminator.

2.2.2. Optional Materials

1. Commercial dot/slot blot apparatus and vacuum source (e.g., Minifold Slotblotter or Dotblotter by Schleicher & Schuell UK Ltd).

2.3. Method

1. Heat DNA samples to 95°C and snap-chill on ice. Add an equal volume of 20× SSC.
2. Place a membrane on top of a sheet of 3MM paper.
3. Spot the samples onto the membrane prewetted in 10× SSC in 2-μL aliquots

allowing to dry between applications. Take care not to allow sample spots to merge into each other. If a dot blot apparatus is used turn on the vacuum source and pipet the samples into the wells of the apparatus.

4. Immerse the membrane in denaturing solution for 5 min.
5. Transfer to neutralizing solution for 1 min.
6. Dry and fix the DNA to the membrane as for Southern blots.
7. Hybridize to a labeled probe as for Southern blots.

3. Colony Blots

3.1. Introduction

Isolation of specific sequences from DNA libraries cloned in either bacteriophage or plasmids can be achieved by plating the library out on agar plates and taking colony lifts from the agar plates. The membranes can then by used in a hybridization reaction using a suitable probe.

3.2. Materials

3.2.1. Specific Materials

1. Hybond N nylon membranes (Amersham Pharmacia Biotech) of the desired diameter (slightly less the diameter of the agar plates the colonies are grown on).
2. 3MM filter paper (Whatman).
3. Sterile needle and blunt-ended forceps.
4. Denaturing solution: 1.5 *M* NaCl, 0.5 *M* NaOH.
5. Neutralizing solution: 1.5 *M* NaCl, 0.5 *M* Tris-HCl, pH 7.2, 0.001 *M* EDTA.
6. 20× SSC: 3 *M* NaCl, 0.3 *M* Na_3 citrate.
7. UV transilluminator.

3.2.2. Optional Materials

1. 10% (w/v) SDS.
2. Oven set at 80°C.

3.3. Method

1. Plate out the bacterial cells or bacteriophage on agar plates and incubate overnight (*see* **Note 1**). Cool to 4°C for at least 30 min.
2. Bend the membrane into a U shape. Place the bottom of the U in contact with the surface of the agar plate and gently fold down so the entire membrane is in contact with the agar plate. Do not move the filter once it is in contact with the surface of the plate as this will result in smearing.
3. Mark the orientation of the membrane with respect to the plate by making three asymmetric holes with a sterile needle. These can be used for orientation of the filter after hybridization.
4. Remove the membrane after 30–60 s with a pair of blunt-ended forceps.

5. Place the membrane face up on a sheet of 3MM paper.
6. A repeat lift can be made from the plate if desired.
7. Once all of the plates have been blotted the DNA is liberated from the colonies by placing the membranes colony side uppermost up on a stack of 3MM paper saturated with denaturing solution (*see* **Note 2**). Leave for 2–5 min.
8. Transfer the membranes (colony side up) to a stack of 3MM paper soaked in neutralizing solution for 3 min.
9. Wash the membranes in 2x SSC with agitation for 2 min to remove cell debris.
10. Place the membranes DNA side up on a pad of dry 3MM paper and allow to dry.
11. Crosslink the DNA to the membrane and hybridize as for Southern blots.

3.4. Notes

1. The colonies should not be allowed to grow too large as they may merge into one another. Aim for a colony density of approx 200 per 83-mm diameter plate. Pre-cooling the plates to 4°C prevents the colonies from smearing when blotted and lowers the amount of agar that adheres to the membrane.
2. The stack of 3MM paper should be moist, but not soaking as this will cause the colonies to diffuse. An optional lysis step can be included before denaturing the DNA by placing the filters on a stack of 3MM paper soaked in 10% SDS for 1–3 min.

4. Northern Blots

4.1. Introduction

RNA must be run on agarose gels under denaturing conditions. Two common methods can be used to achieve this. The glyoxal–dimethyl sulfoxide (DMSO) method and the formaldehyde/formamide method. The latter method is slightly easier and is described here. When working with RNA all glassware should be baked at 180°C overnight and all solutions made up containing 0.2% (v/v) diethylpyrocarbonate (DEPC) and then autoclaved to remove contaminating RNAases. It should be noted that Tris solutions cannot be DEPC-treated and that Tris stock solutions should be made up with DEPC treated water. The Tris stock for RNA work should be taken from a separate container from normal laboratory stocks and should be weighed out only by shaking the Tris out of the container (never use a spatula).

4.2. Specific Materials

1. DEPC for treating all solutions to be used. **Caution:** DEPC is a very dangerous substance and care must be exercised when handling it. Once autoclaved there is no further hazard.
2. 10× MOPS buffer: 0.2 *M* 3-[*N*-Morpholino] propanesulfonic acid, 0.5 *M* Na acetate, pH 7.0; 0.01 *M* EDTA.
3. Formaldehyde: 37% Solution, 12.3 *M*, pH >4.0.

4. Formamide (deionized).
5. 3MM filter paper (Whatman), paper towels, glass or plastic tray and support (a gel casting tray turned upside down), cling film.
6. Hybond N membrane (Amersham Pharmacia Biotech).
7. Denaturing solution: 1.5 *M* NaCl, 0.5 *M* NaOH.
8. Neutralizing solution: 1.5 *M* NaCl, 0.5 *M* Tris-HCl, pH 7.2, 0.001 *M* EDTA.
9. 20× SSC: 3 *M* NaCl, 0.3 *M* Na_3 citrate.
10. UV transilluminator.
11. 100× Denhardt's solution: 10 g of BSA fraction V, 10 g of Ficoll 400, 10 g of PVP in 500 mL of distilled water. Store at –20°C in 10 mL aliquots.
12. 10% (w/v) SDS.
13. 10 mg/mL sheared herring testis DNA. Store at –20°C.

4.3. Method

1. Prepare the Northern gel by dissolving 3 g of agarose in 250 mL of DEPC treated water. Cool to 55°C and add 17.5 mL formaldehyde and 30 mL 10× MOPS buffer (both preheated to 55°C. Cast the gel in an appropriately sized tray in a fume hood.
2. Before loading the samples prerun the gel at 5 V/cm for 5 min in 1× MOPS buffer.
3. Prepare the RNA samples (*see* **Note 1**) for electrophoresis by adding the following to the RNA sample: 5.5 µL of formaldehyde, 15 µL of formamide, 1.5 µL of 10× MOPS buffer and distilled water to a final volume of 30 µL.
4. Denature the sample at 55°C for 15 min and add 3 µL of loading buffer (50% glycerol, 0.25% bromophenol blue in DEPC-treated water). Load onto gel and run (*see* **Note 2**).
5. Place the gel in a tray and cover with DEPC-treated distilled water. Incubate with gentle agitation for 15 min.
6. Remove the water and replace with 10× SSC and shake for 15 min; repeat once.
7. Set up the capillary blot as described for Southern blotting.
8. After transfer is complete dismantle the Northern blot, remove the filter and air-dry. Do not rinse the filter as for Southern blots. Fix the RNA to the membrane by UV crosslinking or baking at 80°C as for Southern blots.
9. Hybridize membrane using the same conditions as for Southern blots.

4.4. Notes

1. The success of any experiments involving RNA depends on the quality of the RNA! The most reliable method to extract RNA from tissue samples is to use a commercially available kit.
2. If possible, recirculate the running buffer.

References

1. Southern, E. M. (1975) Detection of specific sequences among DNA fragments separated by gel electrophoresis. *J. Mol. Biol.* **98,** 503–517.
2. Smith, G. E. and Summers, M. D. (1980) The bidirectional transfer of DNA and RNA to nitrocellulose or diazobenzyloxymethyl-paper. *Ann. Biochem.* **109,** 123–132.

2

End Labeling Procedures

An Overview

Elena Hilario

1. Introduction

The purpose of this chapter is to give an overview of the different end labeling procedures. These protocols have been standardized and optimized by several biotechnology companies and are available in kits. Unless your laboratory is involved in producing large quantities of many different probes (e.g., micrograms of each probe), it is unpractical, and perhaps more expensive, to set up your own protocol. However, a good understanding of the type of probe, the location of the label (3'-, 5'-end or distributed all along the molecule), and the number of labeled nucleotides incorporated should be considered when planning an experiment.

2. Label Location

2.1. 3'-End Labeling

The efficiency of labeling the 3'-end of a DNA molecule depends on the type of enzyme used, the type of DNA molecule (single- or double-stranded), and the length of the 3'-end (recessed, blunt end, or protruding).

DNA polymerase I Klenow fragment (exo$^-$) fills in the 3'-recessed ends of restriction fragments in the presence of radiolabeled nucleotides *(1)*. The number of labeled molecules will depend on the type of labeled nucleotide added and the sequence of the complementary strand. Blunt end fragments can also be labeled by replacing the unlabeled 3'-end nucleotide by a labeled molecule.

From: *Methods in Molecular Biology, vol. 179: Gene Probes: Principles and Protocols*
Edited by: M. Aquino de Muro and R. Rapley © Humana Press Inc., Totowa, NJ

Klenow fragment does not label 3'-protruding ends efficiently. Labeling with Klenow fragment is an appropriate method for producing DNA size markers.

Bacteriophage T4 DNA polymerase synthesizes DNA on a 5'–3' direction. The enzyme also has exonuclease activity 3'–5' but not on the 5'–3' direction. Two steps are involved in labeling probes with bacteriophage T4 DNA polymerase ***(2)***: a replacement reaction using the 3'–5' exonuclease activity of the enzyme in the absence of dNTPs to generate 3'-protruding ends, and then filling in the ends with a mixture of unlabeled and labeled dNTPs. The resulting labeled fragment can be further digested with endonucleases and generate a mixture of probes of different sizes. Alternatively, the 3'-protruding tails of double-stranded DNA, previously digested with endonucleases, are regenerated by the bacteriophage T4 DNA polymerase in the presence of all four dNTPs (including the desired labeled dNTP). In the presence of dNTPs, T4 DNA polymerase 3'–5' exonuclease activity is inhibited and the polymerase activity predominates. However, caution should be taken to avoid long incubations, as the dNTPs could be exhausted, and the 3'–5' exonuclease activity of T4 DNA polymerase will resume and degrade double-stranded DNA as well as single-stranded DNA. Keep in mind that T4 DNA polymerase has a higher rate of 3'–5' exonuclease activity on single-stranded DNA than on double-stranded DNA. The replacement method can be difficult to control; therefore, filling in previously endonuclease digested DNA with 3'- or 5'-protruding ends in the presence of dNTPs is the best alternative when using bacteriophage T4 DNA polymerase for labeling DNA.

Terminal deoxynucleotidyl transferase (TdT) is a template independent DNA polymerase that incorporates dNTPs to the 3'-OH end of single- or double-stranded DNA, and RNA in an irreversible manner ***(3)***. This enzyme is used for the production of synthetic homo- or heteropolymers ***(4)***, for incorporating a homopolymeric tail to any type of DNA 3'-end ***(3,5–8)***, and for incorporating a single nucleotide analog such as [α-^{32}P]cordycepin-5'- triphosphate (Promega) ***(6,9)*** or digoxigenin-11-ddUTP (Roche Molecular Biochemicals) ***(10)***. Terminal dideoxynucletidyl transferase labels 3'-protruding ends more efficiently than blunt ends or 3'-recessed ends. The incorporation of dA or dT residues is more favorable than incorporating dC or dG. The type of method chosen for incorporating labeled nucleotides to the 3'-end of a DNA molecule depends on the required probe sensitivity and specificity. The 3'-end tailing reaction synthesizes highly sensitive probes owing to the addition of several labeled molecules, but the specificity decreases owing to unspecific binding of the added nucleotide tail. This inconvenience can be solved by changing the stringency conditions. If probe specificity is the priority, 3'-end labeling of the DNA molecule should be performed.

2.2. 5'-End Labeling

There are three ways of labeling DNA molecules at the 5'-end: enzymatic, chemical, or combined methods. A brief description of each method is given.

2.2.1. Enzymatic Methods

The bacteriophage T4 polynucleotide kinase catalyzes two reactions: forward and exchange. In the forward reaction, the enzyme transfers the γ phosphate of [γ-^{32}P]ATP to the 5'-hydroxy group of a DNA molecule (oligonucleotides or nucleoside 3'-monophosphates) or RNA, previously dephosphorylated with alkaline phosphatase. In the exchange reaction, T4 polynucleotide kinase transfers the 5'-terminal phosphate group of the DNA molecule to ADP. Then, the enzyme transfers the γ phosphate of [γ-^{32}P]ATP to the 5'-hydroxy group of a DNA molecule. The forward and the exchange reactions depend on the amount of ATP available ***(13,14)***. The wild-type bacteriophage T4 polynucleotide kinase has 3'-phosphatase activity ***(11)***; however, this unwanted property has been engineered and a mutant T4 polynucleotide kinase 3'-phosphatase minus enzyme is now available (MBI Fermentas, Roche Molecular Biochemicals) ***(12)***. T4 polynucleotide kinase preferentially labels protruding 5'-ends over blunt ends or recessed 5'-ends, but in the presence of polyethylenglycol 8000 the reaction conditions for labeling blunt ends or recessed 5'-ends can improve ***(13)***. Precaution should be taken in avoiding ammonium and phosphate ions during any purification procedure, as T4 polynucleotide kinase is strongly inhibited by these ions ***(13,14)***.

2.2.2. Chemical Methods

This approach is suitable for synthetic oligonucleotides with a modified 5'-end. Terminal amino function is incorporated to the 5'-end after synthesizing the oligonucleotide by adding a phosphoroamidite group. After cleavage from the synthesis support and activation, a digoxigenin molecule is covalently linked (Roche Molecular Biochemicals) ***(15)***. Large quantities of oligonucleotide can be labeled per reaction. The 3'-end remains undisturbed and available for primer extension.

2.2.3. Combined Method

Two methods have been developed by Promega to incorporate nonradioactive labels to the 5'- or/and 3'-end of unmodified or modified oligonucleotides. The T4 polynucleotide kinase incorporates a thiophosphate from adenosine-5'-*O*-(3-thiotriphosphate) to an unmodified oligonucleotide. Then, the activated thiol group of the oligonucleotide reacts with a maleimide modified hapten (fluorescein or biotin), leaving the 3'-end of the molecule unaltered

(FluoroAmpJ T4 Kinase System, Promega). The alternative method generates alkaline phosphatase conjugates at the 5'- or 3'-end of unmodified or amino modified oligonucleotides (LIGHTSMITH® II System, Promega Corp.). If the starting oligonucleotide is not modified, terminal dideoxynucleotidyl transferase incorporates an amino modified ATP to the 3'-end before the activation and conjugation of the enzyme hapten.

3. Purification of Labeled Probes

Radiolabeled probes are usually not purified after synthesis; however, if the incorporation yield is low, removing the unincorporated label might help to avoid a high background noise.

There are four methods to purify labeled probes: spin column chromatography, membrane filtration, adsorption to silica gel membranes, and ethanol precipitation. Gel size exclusion properties in spin column chromatography are not the same as in flow-dependent fractionation. In spin column chromatography, *g* force applied to the column and the centrifugation time are important factors during the exclusion process. Using prepacked and equilibrated columns prevents dilution of the applied sample. Reproducibility during purification is achieved with commercially prepacked columns (e.g., Amersham Pharmacia, Roche Molecular Biochemicals). Specific resins are used for particular purification procedures such as buffer exchange and desalting, and removal of excess primers or free nucleotides.

Membrane filtration is a fast and reliable way of removing excess label, exchanging buffers, and concentrating a sample. Microcentrifuge devices are commercially available with different cut off ranges (e.g., Microcon®, Millipore). This method is appropriate to remove primers, linkers, labeled nucleotides and desalting samples; however, modifying enzymes are usually retained together with the labeled DNA molecule. Therefore, membrane filtration is recommended mainly for chemical labeling methods.

Adsorption of DNA molecules of certain size ranges to silica-gel membrane occurs at high ionic strength and is eluted at lower ionic strength (Qiagen, Germany). No ethanol precipitation is required. Silica-gel membranes assembled on spin columns overcome the problems associated with silica-gel slurries (low yields, slurry carried over with eluted DNA, etc.). Free labeled nucleotides, modifying enzymes, reaction buffers, and other components of the labeling reaction mixtures are easily removed by this method. Recovery of oligonucleotides (17–40 bases long) and double- or single-stranded DNA fragments up to 10 kb long is feasible. This is the most efficient way to clean up any modification reaction.

Ethanol precipitation with ammonium or sodium acetate can be performed for most labeling procedures ***(13,14)***; however, for Digoxigenin-labeled probes, lithium chloride (final concentration 0.4 *M*) should be used instead of sodium acetate. Oligonucleotides and low concentration of labeled probes are easily precipitated with carrier molecules such as glycogen (final concentration 0.4 mg/µL). Alkaline conjugates and fluorescein- or biotin-labeled probes require a combination of spin column chromatography and ethanol precipitation. The storage temperature for most probes is –20°C, or temporarily 4°C. Specific storing buffers are recommended for each method, and special care should be taken regarding pH conditions, light exposure, stabilizers (e.g., glycerol), and half-life of the probe. Storing the synthesized probe in small aliquots prevents degradation by repeated freeze–thaw cycles, and the possibility of accidental cross contamination with other probes.

References

1. Derbyshire, V., Freemont, P. S., Sanderson, M. R., Beese, L., Friedman, J. M., Joyce, C. M., and Steitz, T. A. (1988) Genetic and crystallographic studies of the 3',5'-exonucleolytic site of DNA polymerase I. *Science* **240,** 199–201.
2. Challberg, M. D. and Englund, P. T. (1980) Specific labeling of 3' termini with T4 DNA polymerase. *Methods Enzymol.* **65,** 39–43.
3. Deng, G. and Wu, R. (1983) Terminal transferase: use of the tailing of DNA and for *in vitro* mutagenesis. *Methods Enzymol.* **100,** 96–115.
4. Chang, L. M. and Bollum, F. J. (1986) Molecular biology of terminal transferase. *CRC Crit. Rev. Biochem.* **21,** 27–52.
5. Eschenfeldt, W. H., Puskas, R. S., and Berger, S. L. (1987) Homopolymeric tailing. *Methods Enzymol.* **152,** 337–342.
6. Tu, C. and Cohen, S. (1980) 3'-End labeling of DNA with [α-^{32}P]cordycepin-- 5'-triphosphate. *Gene* **10,** 177–183.
7. Vincent, C., Tchen, P., Cohen-Solal, M., and Kourilsky, P. (1982) Synthesis of 8-(2-4-dinitrophenyl 2-6 aminohexyl) amino-adenosine 5' triphosphate: biological properties and potential uses. *Nucleic Acids Res.* **10,** 6787–6796.
8. Kumar, A., Tchen, P., Roullet, F., and Cohen, J. (1988) Non radioactive labeling of synthetic oligonucleotide probes with terminal deoxynucleotidyl transferase. *Anal. Biochem.* **169,** 376–382.
9. Brown, T. A., ed. (1998) *Molecular Biology Labfax, I. Recombinant DNA*, 2nd edit., Academic Press, San Diego.
10. Schmitz, G. G., Walter, T., Seibl, R., and Kessler, C. (1991) Nonradioactive labeling of oligonucleotides *in vitro* with the hapten digoxigenin by tailing with terminal transferase. *Anal. Biochem.* **192,** 222–231.
11. Cameron, V. and Uhlenbeck, O. C. (1977) 3'-Phosphatase activity in T4 polynucleotide kinase. *Biochemistry* **16,** 5120–5126.
12. Cameron, V., Soltis, D., and Uhlenbeck, O. C. (1978) Polynucleotide kinase from a T4 mutant which lacks the 3' phosphatase activity. *Nucleic Acids Res.* **5,** 825–833.
13. Sambrook, J., Fritch, E. F., and Maniatis, T. (1989) *Molecular Cloning: A Labo-*

ratory Manual, 2nd edit., Cold Spring Harbor Laboratory, Cold Spring Harbor, New York.

14. Ausubel, F. M., Brent, R., Kingston, R. E., Moore, D. D., Seidman, J. G., Smith, J. A., and Struhl, K., eds. (1994) *Current Protocols in Molecular Biology*. John Wiley & Sons, New York.
15. Zischler, H., Nanda, I., Schafer, R., Schmid, M., and Epplen, J. T. (1989) Digoxigenated oligonucleotide probes specific for simple repeats in DNA fingerprinting and hybridization *in situ*. *Hum. Genet.* **82,** 227–233.

3

Photobiotin Labeling

Elena Hilario

1. Introduction

Biotin is a small vitamin found in tissue and blood and is synthesized by intestinal bacteria. Biotin functions as a prosthetic group for several carboxylases and as a CO_2 carrier. The molecule consists of a ring system covalently linked to the enzyme by a valerate side chain acting as a flexible arm. Avidin and streptavidin are two proteins that strongly bind biotin. Avidins are tetrameric proteins; each subunit has a molecular mass of 13,000–16,000 Da. Avidin is found in chicken egg whites, and a lower affinity variant is found in the yolk. Streptavidin is found in the fungus *Streptomyces avidinii.* Although the dissociation constant of avidin for biotin is higher than that of streptavidin for biotin (10^{-15} *M* and 10^{-14} *M*, respectively), the nonspecific adsorption of streptavidin to nucleic acids and negatively charged cell membranes is preferred over avidin. This is due to the low isoelectric point of streptavidin (pI 5–6), compared to pI 10 for avidin. The binding characteristics and stability of avidin–biotin and streptavidin–biotin complexes have been extensively explored and applied to develop numerous methods in immunology and molecular biology.

There are three methods for labeling DNA molecules with biotin: chemical, enzymatic, and photolabeling reactions. The type of method to be used depends on the amount of DNA available and the number of biotin molecules to be incorporated. Several commercial kits are available for DNA biotinylation. The *cis*-platinum Chem-Link reagent (Kreatech Biotechnology, B.V., The Netherlands) binds to the N7 position of guanosine and adenosine bases at 85°C, forming a stable biotin–Pt complex. This chemical reaction might denature and fragment the target DNA; however, this method is suitable for blot hybrid-

From: *Methods in Molecular Biology, vol. 179: Gene Probes: Principles and Protocols*
Edited by: M. Aquino de Muro and R. Rapley © Humana Press Inc., Totowa, NJ

ization. Biotin hydrazide (five carbon atoms in space arm) and Biocytin-hydrazide (10 carbon atoms in space arm) (Pierce, Perstorp Biotec) label cytosine residues in DNA and RNA via a bisulfite-catalyzed transamination reaction ***(1)***. Biotin analogs of dTTP with long spacer arms (16 carbon atoms) can be used for incorporating biotin enzymatically: by 3' end-labeling with terminal transferase and biotin-16-ddUTP, or by replacing dTTP with biotin-16-dUTP in nick translation, random priming, or polymerase chain reaction (PCR) amplifications. All the methods described (except PCR biotin labeling) require considerable amounts of starting DNA substrate, which must be free of any oligonucleotides, buffers, enzymes, additives, etc. Forster and co-workers first described the method to synthesize photobiotin acetate and bind it to nucleic acids. A molecule of photobiotin is formed by a biotin molecule bound to a linker arm of nine carbons with a positively charged tertiary amino group at the center and a photoreactive arylazide group. The reaction takes place under strong visible light or under ultraviolet light. The arylazide molecule is converted to an extremely reactive arylnitrene group that binds with the nucleic acid; however, the nature of the linkages is unknown. The spacer arm is long enough to allow the interaction of biotin and streptavidin without interference ***(2)***.

What is the advantage of using photobiotin over the chemical or enzymatic biotin labeling methods? Photobiotinylation is a simple method for labeling DNA because the reaction occurs in water and is terminated by mild alkalinization. Although ultraviolet light is used to photoactivate the molecule, the DNA damage is minimum. It does not interfere in hybridizing the biotinylated probe to the target DNA molecules in any further protocol, including delicate procedures such as subtraction hybridization ***(3,4)***.

Photobiotin acetate can be synthesized in the laboratory ***(5)***; however, it is also available from Pierce Chemical (Rockford, IL) and Sigma-Aldrich (St. Louis, MO).

2. Materials

1. 1 mg of photobiotin acetate.
2. Insect vials with flat bottom, without caps.
3. Scintillation vials without caps and a styrofoam floating boat.
4. Tray with ice water.
5. Darkroom with red safety light (e.g., KODAK safety light filter GBX2).
6. UV transilluminator lamp (MacroVue UV-25 transilluminator 302 nm Hoefer, set in HIGH 254 nm, 9000 $\mu W/cm^2$).
7. 1 *M* Tris, pH 9.0.
8. Tris–EDTA-saturated *n*-butanol:10 m*M* Tris-HCl, pH 8.0, 1 m*M* EDTA.
9. 5 *M* Potassium acetate, pH 7.0.

10. Isopropanol.
11. 70% Ethanol.
12. Appropriate final resuspension buffer at pH 7.5–8.0.
13. 1% Agarose gel for DNA electrophoresis.

3. Method: Photobiotin Labeling

Steps 1–7 must be done in a darkroom, with a red safety light.

1. Prepare a 1 µg/µL stock solution of photobiotin acetate in deionized water. Store in a desiccator, at –20°C, protected from light. The solution is stable for about 6 mo and should be bright orange when fresh (*see* **Note 1**).
2. Add 50 µL of photobiotin solution to the 50-µL sample of sized sheared DNA (~1 µg/µL in deionized water). Mix very well.
3. Assemble a "thermus device" by inserting an insect vial inside a scintillation vial with ice water. Transfer the sample to the bottom of the insect vial.
4. Place the insect/scintillation vials in a floating boat on the ice water tray.
5. Place the icy water tray under the UV transilluminator lamp (*see* **Note 2**). The distance between the bottom of the insect vial to the surface of the transilluminator should be 10 cm.
6. Irradiate the sample for 30 min. Mix the solution by gently tapping the vial every 5 or 10 min to have a uniform exposure to UV light. After 30 min the mixture turns orange-brown.
7. Stop the reaction by adding 10 µL of 1 *M* Tris-HCl, pH 9.0.
8. Extract the unreacted photobiotin with an equal volume of Tris–EDTA-saturated *n*-butanol, 3× or until organic phase is clear.
9. Precipitate the sample with 1/10th of the volume of 5 *M* potassium acetate, pH 7.0, and 1 volume of isopropanol and store at –20°C overnight. Centrifuge the sample in a microcentrifuge at maximum speed for 20 min. The color of the small pellet should be orange-brown. Wash 2× with 500 µL 70% ethanol, and centrifuge again. Dry the pellet protected from direct light. Resuspend the pellet in 30 µL of the appropriate resuspension buffer at pH 7.5–8.0 (*see* **Note 3**). If necessary, heat up the sample at 65°C for 5 min for complete resuspension. Vortex-mix briefly and centrifuge to collect all the sample. The approximate concentration should be 1 µg/µL.
10. Run 5 µL of biotinylated sample, and 5 µL of nonbiotinylated DNA in a 1% agarose gel. The biotinylated sample should run slower than the unlabeled DNA. A minimum amount of sample might stay inside the well. A recovery of 70–75% of biotinylated DNA is expected for this procedure. Store the biotinylated DNA sample at –20°C. Avoid freeze–thaw cycles by storing appropriate aliquots of the probe. The biotinylated DNA is stable for up to a year (*see* **Note 4**).

4. Notes

1. Photobiotinylation is a simple technique to label DNA. The photochemical reaction takes place in an aqueous phase, which simplifies the recovery of the labeled DNA. Attention must be taken in storing the photobiotin solutions at –20°C in the dark, and try to use up the whole stock within 6–12 mo. Ordering the smallest quantity available from the supplier company is preferred.
2. McInnes and colleagues recommend a white light lamp instead of ultraviolet light *(5)*. Ten to twenty minutes are enough to complete the labeling reaction. Examples of commercial lamps available are: I-Line Multi-Vapor® Lamp (General Electric 10687 325W or 43817 400W), or sun lamps (GE Model RSM 250 W).
3. The chemical composition of the final resuspension buffer varies according to the protocol requiring the biotinylated DNA. High concentrations of NaCl up to 1 *M* or detergents such as dodecyl sodium sulfate (0.1%) have no effect on the labeled DNA. The type of buffer used is not critical, as long as it is within pH 7–8.5.
4. To avoid incomplete labeling of your DNA sample, do not scale up the labeling reaction volume. If more than 50 μg of DNA need to be labeled, prepare as many vials containing 50 μL each, as needed.
5. **General comment:** The most convenient retrieving molecule to capture the biotinylated DNA is streptavidin. Streptavidin is usually linked to a reporter enzyme such as alkaline phosphatase which produces a detectable colored product in standard methods for Southern and Northern blot analysis. Other ways to capture streptavidin–biotinylated DNA complexes from aqueous solutions are columns coupled with streptavidin or avidin, or phenol–chloroform extractions.

References

1. Reisfield, A., Rosenberg, J. M., Bayer, E. A., and Wilchek, M. (1987) Nonradioactive hybridization probes prepared by the reaction of Biotin-Hydrazide with DNA. *Biochem. Biophys. Res. Commun.* **142,** 519–526.
2. Forster, A. C., McInnes, J. L., Skingle, D. C., and Symons, R. H. (1985) Nonradioactive hybridization probes prepared by the chemical labeling of DNA and RNA with a novel reagent, photobiotin. *Nucleic Acids Res.* **13,** 745–761.
3. Sun, T-P., Goodman, H. M., and Ausubel, F. M. (1992) Cloning the *Arabidopsis* GA1 locus by genomic subtraction. *Plant Cell* **4,** 119–128.
4. Straus, D. and Ausubel, F. M. (1990) Genomic subtraction for cloning DNA corresponding to deletion mutations. *Proc. Natl. Acad. Sci. USA* **87,** 1889–1893.
5. McInnes, J. L., Forster, A. C., Skingle, D. C., and Symons, R. H. (1990) Preparation and uses of photobiotin. *Methods Enzymol.* **184,** 588–600.

4

Nick Translation

Alex Reid

1. Introduction

Nick translation was the first method devised for the in vitro labeling of DNA *(1)*. During the reaction the DNA to be labeled is nicked by DNase I yielding a free 3' hydroxyl end. DNA polymerase I then adds a new nucleotide to this end. The 5'–3' exonuclease activity of the polymerase then moves the "nick" along the strand in the 3' direction. The addition of a radioactively labeled nucleotide to the reaction results in probes that can be used in hybridization reactions to DNA immobilized on Southern blots, colony lifts, etc.

2. Materials

2.1 Specific Materials

1. Nick Translation System (Promega), including enzyme mix, unlabeled dNTP solutions, reaction buffer (10×), stop solution and nuclease free water.
2. [α-^{32}P]dNTP (normally dCTP) at 3000 Ci/mmol; 10 µCi/µL. **Important:** Radioactively labeled material should be only handled in a designated laboratory following safety guidelines. Work should be carried out behind Perspex screens and all solid and liquid waste disposed in designated containers.
3. Sephadex G-50, equilibrated in 1× TE buffer (10 m*M* Tris-HCl, pH 8.0, 1 m*M* EDTA).
4. Glass wool.

2.2 Optional Materials

1. 0.2 *M* EDTA, pH 8.0.
2. 0.5 *M* Sodium phosphate, pH 6.8 (*see* **Note 1**).
3. Whatman DE 81 2.3-cm circular filters, scintillation counter. These are only required if the percentage incorporation of label is needed to be known.

From: *Methods in Molecular Biology, vol. 179: Gene Probes: Principles and Protocols*
Edited by: M. Aquino de Muro and R. Rapley © Humana Press Inc., Totowa, NJ

3. Method

1. Place kit components (except enzyme mix which should be kept at –20°C until required) at room temperature until thawed and transfer to ice.
2. On ice add the following to a sterile 1.5-mL of centrifuge tube, 10 μL nucleotide mix (*see* **Note 2**), 5 μL of 10× nick translation buffer, 1 μg of DNA, 70 μCi of [α-^{32}P] dCTP, 5 μL of nick translation enzyme mix (*see* **Note 3**), nuclease-free water to a final volume of 50 μL.
3. Incubate at 15°C for 60 min (*see* **Note 4**).
4. Add 5 μL of stop solution.
5. To calculate the percentage incorporation of radioactivity and the specific activity of the probe remove 1 μL of the reaction and add 99 μL of 0.2 *M* EDTA, pH 8.0. Spot 3 μL of the dilution onto two Whatman DE81 2.3-cm circular filters. Dry the filters under a desk lamp for 5 min. Wash one filter in 50 mL 0.5 *M* sodium phosphate, pH 6.8, twice for 5 min each and dry the filter under the desk lamp for 5 mins. Place the filters in scintillation vials and read on the tritium channel (*see* **Note 5**).
6. The unincorporated nucleotides can be separated from the reaction by centrifuging the reaction through a Sephadex G-50 column.
7. Plug the bottom of a 1-mL syringe with sterile glass wool to the 0.1-mL mark. This is best done by packing the glass wool down with the syringe barrel.
8. Fill the syringe to the top with Sephadex G-50 equilibrated in 1× TE buffer. Place the syringe in a 15-mL disposable tube and centrifuge at 1600*g* for 4 min in a swinging bucket rotor. The resin should pack down to the 0.9-mL mark on the syringe. If not add more Sephadex G-50 and recentrifuge. Add 100μL 1× TE buffer to the top of the column and centrifuge as before. Transfer the column to a fresh 15-mL centrifuge tube.
9. Add 45 μL of 1× TE buffer to the nick translation reaction to bring the volume up to 100 μL and add to the top of the column. Centrifuge the column at 1600*g* for 4 min and discard the column in the radioactive waste.
10. Transfer the elutate to a 1.5-mL Safe-Lock Eppendorf tube. Discard the centrifuge tube in the radioactive waste.
11. It is advisable to use labeled probes immediately as the radioactive decay will damage the DNA over extended periods of time. If the probe is needed at a later date they can be stored at –20°C in a lead pot.
12. Before use in a hybridization reaction denature the labeled DNA by heating to 95–100°C for 5 min and chill on ice to prevent renaturation. The use of a Safe-Lock Eppendorf tube will prevent the cap opening during heating.

4. Notes

1. 0.5 *M* Sodium phosphate, pH 6.8: 47.25 g of NaH_2PO_4, 22.35 g of Na_2HPO_4 in 1 L of distilled water.
2. The unlabeled nucleotide mix is prepared by combining equal volumes of the three dNTPs that are not labeled, for example, dATP, dGTP and dTTP if the labeled nucleotide to be used is [α-^{32}P]dCTP.

3. Although no details of the enzymes present in the Promega kit are given standard reaction conditions contain pancreatic DNase I (10 ng/mL) and 2.5 U of *E. coli* DNA polymerase I.
4. At temperatures higher than 20°C there will be a significant quantity of "snapback" DNA produced by the *E. coli* DNA polymerase I where the enzyme starts copying the newly synthesized strand instead of the template DNA.
5. This method detects the Cerenkov radiation produced by the decay of ^{32}P and does not give an absolute number of counts. The readings are however, proportional from sample to sample. The percent incorporation can be calculated by dividing the counts from the washed filter by the counts from the unwashed filter and multiplying by 100. The percentage incorporation should be above 60%. The specific activity of the probe is obtained by dividing the number of counts on the washed filter by the amount of DNA added to the labeling reaction and multiplying by 33.3 (3 µL of a 1:100 dilution) and the multiplying by 55 (the reaction volume). Specific activity is expressed as cpm/µg and a good reaction will yield around 10^8 cpm/µg. Although it is not essential to measure the percent incorporation and specific activity of probes before their use, it does enable the monitoring of reactions from one experiment to another.

References

1. Kelly, R. B., Cozzarelli, N. R., Deutscher, M. P., Lehman, I. R., and Kornberg, A. (1970) Enzymatic synthesis of deoxyribonucleic acid. XXXII. Replication of duplex deoxyribonucleic acid by polymerase at a single strand break. *J. Biol. Chem.* **245,** 39–43.

5

Random Primed Labeling

Alex Reid

1. Introduction

Labeling of DNA by nick translation has three major drawbacks: the time taken to perform the reaction (at least 1 h), the temperature sensitivity of the reaction, and the low specific activity of the probes generated. Random primed labeling developed by Feinberg and Vogelstein ***(1,2)*** solves all of these problems. The technique uses short random sequence hexanucleotides (in the original method) which prime the denatured target DNA at numerous sites. The Klenow fragment of DNA polymerase I is then used to synthesize new strands of DNA from these primed sites. The addition of a radioactive nucleotide results in a labeled probe suitable for use in Southern hybridizations, etc. Small amounts of starting material are required and the 10-min reaction results in probes labeled to a high specific activity.

2. Materials

2.1 Specific Materials

1. Megaprime DNA labeling system (Amersham Pharmacia Biotech) including primer solution (random nonamers), labeling buffer (includes dATP, dGTP and dTTP), enzyme solution (DNA polymerase I Klenow fragment) [α-^{32}P]dCTP at 3000 Ci/mmol; 10 µCi/µL. **Important:** Radioactively labeled material should be handled only in a designated laboratory following safety guidelines. Work should be carried out behind Perspex screens and all solid and liquid waste disposed in designated containers.
2. Sephadex G-50, equilibrated in 1× TE buffer (10 m*M* Tris-HCl, pH 8.0; 1 m*M* EDTA).
3. Glass wool.

From: *Methods in Molecular Biology, vol. 179: Gene Probes: Principles and Protocols*
Edited by: M. Aquino de Muro and R. Rapley © Humana Press Inc., Totowa, NJ

2.2 Optional Materials

1. 0.2 *M* EDTA, pH 8.0.
2. 0.5 *M* Sodium phosphate, pH 6.8 (*see* **Note 1**).
3. Whatman DE 81 2.3-cm circular filters, scintillation counter. These are required only if the percentage incorporation of label needs to be known.

3. Method

1. Place kit components (except enzyme mix which should be kept at –20°C until required) at room temperature until thawed and transfer to ice.
2. Add the following to a sterile 1.5-mL centrifuge tube at room temperature, 25 ng of template DNA (*see* **Note 1**), 5 µL of primer solution, and distilled water to give a final reaction volume of 50 µL. Denature by heating at 95–100°C for 5 min. Centrifuge briefly to collect the contents at the bottom of the tube.
3. At room temperature add the following to the tube 5 µL labeling buffer (including unlabeled nucleotides), 2 µL of [α-^{32}P]dCTP (10 µCi/µL), 2 µL. Mix gently and centrifuge briefly.
4. Incubate reaction at 37°C for 10 min.
5. Stop the reaction by adding 0.2 *M* EDTA.
6. To remove unincorporated nucleotides and to calculate the percent incorporation and specific activity (*see* **Note 2**) follow the procedures outlined in the Nick Translation protocol.

4. Notes

1. Dilute the template DNA to a concentration of 25 ng/mL in either distilled water or 1× TE buffer (10 m*M* Tris-HCl, pH 8.0, 1 m*M* EDTA). Bands excised from low melting point gels run with TAE buffer can also be labeled using the megaprime kit.
   Place the excised band (with as little agarose as possible) in a 1.5-mL Eppendorf tube and add distilled water to 3-mL/g of gel. Place the tube in a boiling water bath for 5 min. Add an appropriate volume for 25 ng of template DNA to the reaction (not to exceed 25 µL in a 50 µL reaction volume). Increase the incubation period for excised bands to 30 min.
2. The specific activity of probes labeled using the random primed method are generally of a higher specific activity (>10^9 cpm/µg) than those labeled by nick translation.

References

1. Feinberg, A. P. and Vogelstein, B. (1983) A technique for radiolabeling DNA restriction endonuclease fragments to high specific activity. *Ann. Biochem.* **132,** 6–13.
2. Feinberg, A. P. and Vogelstein, B. (1984) Addendum: A technique for radiolabeling DNA restriction endonuclease fragments to high specific activity. *Ann. Biochem.* **137,** 266–267.

6

Design and Evaluation of 16S rRNA-Targeted Oligonucleotide Probes for Fluorescence *In Situ* Hybridization

Philip Hugenholtz, Gene W. Tyson, and Linda L. Blackall

1. Introduction

Fluorescence *in situ* hybridization (FISH) of whole cells using 16S rRNA-targeted oligonucleotide probes is a powerful technique with which to evaluate the phylogenetic identity, morphology, number, and spatial arrangements of microorganisms in environmental settings *(1)*. Probes can be designed to specifically target narrow to broad phylogenetic groups (from species to domain) by virtue of variable evolutionary conservation within the 16S rRNA molecule *(2)*. The major steps in probe design are identifying short regions (usually 15–25 nucleotides in length) in a sequence alignment unique to the target group of interest, centralizing mismatches to nontarget organisms (where possible), and modifying the sequence to meet probe design criteria such as a minimum melting temperature.

The FISH method involves application of oligonucleotide probes to permeablized whole microbial cells. The probes enter the cells and specifically hybridize to their complementary target sequence in the ribosomes. If no target sequence is present in the cells ribosomes, probes are unable to hybridize and unbound probe is removed by a subsequent wash step. Hence only specifically targeted cells retain the probes under appropriate stringency conditions in the hybridization and wash steps. Probes are typically 5' end-labeled with fluorochrome reporters such as fluorescein or sulfoindocyanine (Cy3, Cy5) dyes and cells containing hybridized probes can be directly observed under epifluorescence microscopy owing to the natural amplification of the fluorescent signal by large numbers of ribosomes in any given target cell. An advantage is

From: *Methods in Molecular Biology, vol. 179: Gene Probes: Principles and Protocols*
Edited by: M. Aquino de Muro and R. Rapley © Humana Press Inc., Totowa, NJ

that multiple probes with varying target specificity can be used in the same preparation providing they are labeled with clearly distinguishable fluorochromes (i.e., well separated emission wavelengths). For instance, up to seven phylogenetically distinct groups of organisms can be visualized using a combination of three fluorochromes ***(3)***.

The method was first applied using radioactive reporters ***(4)***, which provided only limited microscopic resolution of cells and required an extra step (visualization by microautoradiography). The first demonstration of the FISH method in its modern form was by DeLong and co-workers ***(5)***, using a simple artificial microbial consortium. Subsequently, FISH has been applied in a variety of natural and artificial ecosystems confirming the great utility of the method ***(1)***. However, there are also a number of limitations associated with the method such as poor cell permeability, ribosome accessibility, and content and sample autofluorescence ***(1)***. FISH probes have been designed mainly to target 16S rRNAs but also 23S rRNAs ***(1)***. The aim of this chapter is to specifically address the design and evaluation of 16S rRNA targeted probes used in the FISH method.

2. Materials

2.1. Probe Design

1. Sequence database and phylogeny software: ARB, freeware available from the Department of Microbiology, Technical University, Munich: *http://www.arb-home.de/*. ARB runs on a UNIX platform, LINUX for PCs is recommended.
2. On-line public database search program: Basic local alignment search tool (BLAST): *http://www.ncbi.nlm.nih.gov/BLAST/*
3. On-line oligo parameter calculation programs:
   Biopolymer calculator: *http://www.ncbi.nlm.nih.gov/ blast/blast.cgi*
   Primer calculator: *http://www.williamstone. com/primers/calculator/*
4. *E. coli* probe accessibility table:
   http://aem.asm.org/cgi/content-nw/full/64/12/4973/T1

2.2. Probe Evaluation

1. On-line probe synthesis companies: Interactiva (*http://www.interactiva.de/*), Genset oligos (*http://www.gensetoligos.com/*).
2. Teflon-coated glass slides with 8–12 individual wells (*http://www.superior.de/*).
3. Sterile milli-Q water.
4. 1× phosphate-buffered saline (PBS): 130 m*M* NaCl, 10 m*M* sodium phosphate buffer, pH 7.2. For pH 7.2, the ratios of disodium/sodium phosphates must be 2.57:1.
5. 0.2-μm membrane filters.

6. 2 *M* NaOH, 2 *M* HCl.
7. Fixative solutions: 4% paraformaldehyde. Heat 65 mL of milli-Q water to 60°C. Add 4 g of paraformaldehyde. Add a few drops of 2 *M* NaOH solution and stir rapidly until the solution has nearly clarified (approx 1–2 min). Remove from the heat source and add 33 mL of 3× PBS. Adjust pH to 7.2 with 2 *M* HCl. Remove any remaining crystals by sterile filtration (0.2 μm). Quickly cool and store in 2-mL aliquots at –20°C.
8. 50%, 80%, 98%, and 100% ethanol.
9. Hybridization oven (e.g., Hybaid).
10. Two-milliter microcentrifuge tubes (sterile).
11. Hybridization and wash buffer ingredients: 5 *M* NaCl, 1 *M* Tris-HCl, 10% sodium dodecyl sulfate (SDS). All ingredients should be autoclaved except the SDS solution which should be prepared in sterile milli-Q water.
12. 100% Formamide in 2 mL aliquots.
13. 50-mL polypropylene screw-capped tube—one slide per tube for hybridization.
14. Paper towel.
15. FISH probes at working concentration (50 ng/μL).
16. 48°C water bath.
17. Antifading solution (e.g., Citifluor).
18. Large coverslips.
19. Epifluorscence or confocal laser scanning microscope.

3. Methods

Before embarking on the relatively lengthy process of FISH probe design and evaluation, it is worthwhile checking the literature to ensure a suitable probe does not already exist for your organism(s) of interest (target organism or group). This is a useful exercise because even well-designed oligonucleotides will not always be successful as FISH probes for reasons independent of probe design. Many on-line resources are available to search for existing FISH probes, including literature databases such as Web of Science (*http://wos.isiglobalnet.com/*, requires license) and PubMed (*http://www.ncbi.nlm.nih.gov/Entrez/*), which can be searched by keyword. A good combination of keywords to use is the name of your organism(s) of interest and the word "probe." Also, a number of rRNA probe databases are available on-line including Oligo Retrieval System (ORS; *http://soul.mikro.biologie.tu-muenchen.de/ORS/*) and Oligonucleotide Probe Database (OPD; *http://www.cme.msu.edu/OPD/*). These databases provide details of probes optimized for a number of applications, including FISH. The ORS database can be searched by keyword; however, the OPD database can be searched only by browsing through "Target Nucleic Acids and Data."

3.1 Probe Design

The objective of probe design is to select an oligonucleotide sequence completely specific (complementary) to a region of the target sequences which has at least one mismatch to the same region in all other (nontarget) sequences. A common rule of thumb is to centralize the mismatch or mismatches in the nontarget sequences to maximize the destabilizing effect of the mismatch ***(6)***. Initial probe design can be performed manually or by using computer programs, such as the freeware program ARB, which has a probe design function. A step-by-step description of probe design using ARB is given below; however, it is beyond the scope of this chapter to fully describe how to use the ARB program, and help is provided within the program. Pull down menus in ARB, used to perform the listed tasks, are indicated following a colon.

1. Unmark all sequences in the ARB database before proceeding: S*p*ecies/*U*nmark all Species
2. Mark sequences in the ARB database for which you wish to design a probe (target sequences): click on MARK button at top of left hand vertical command column and then click on individual sequences required in main window (*see* **Note 1**).
3. Open probe design window: *E*tc/Probe *F*unctions/Probe *D*esign…
4. Select a PT_SERVER (*see* **Note 2**), and define parameters. We usually only adjust the minimum percentage of group hits (by setting Min group hits (%) to 100), and maximum number of nongroup hits (*see* **Note 3**). The default target string length is 18 nt.
5. Click on GO. Results appear in PD RESULT window. Note you will not always get a result (*see* **Note 4**). Potential target sequences and associated parameters, such as length, location (*E. coli* number), G+C content, and melting temperature are displayed in the PD RESULT window. Often several close variations of a potential probe site (e.g., A) are displayed, shifted one or more nucleotides upstream (denoted as A+, A++, etc.) or downstream (A-, A−−, etc.) relative to A. Highlight a target sequence of interest for further analysis by clicking on it.
6. Open probe match window: *E*tc/Probe *F*unctions/Probe *M*atch…A valuable feature of ARB is that fields are linked between windows. Therefore the highlighted target sequence in the PD RESULT window will appear in the Target String field of the PROBE MATCH window ready for further analysis.
7. Select a PT_SERVER (should be same PT-SERVER selected previously), and set Search depth to SEARCH UP TO NULL MISMATCHES. Click on MATCH. Sequences in the database with no mismatches to the target string will appear in the match window, and should more or less comprise the initial selection of target sequences used for the probe design confirming the putative probe specificity. A positional match in the target string is indicated by = in the match window.
8. Change the Search depth to one or more mismatches, and click on MATCH. This time nontarget sequences with one or more mismatches to the target string will

appear in the match window below the sequences with an exact match to the target string. Positional mismatches are indicated by the nucleotide that does not form a canonical pairing with its target string complement. Strong mismatches (A:A, C:C, G:G, U/T:U/T, A:C, C:U/T) are shown in uppercase and weak mismatches (a:g, g:u/t) are shown in lowercase. A useful probe will contain one or more mismatches to non-target sequences, ideally located in the middle of the target string. It may be possible to centralize the mismatch(es) by checking variations (A, A-, A+) proposed by the probe design program. If a nontarget sequence contains only a single weak mismatch to the target string, it may be necessary to design a competitor probe (*see* **Note 5**).

9. Once target strings are identified that have at least one mismatch to all nontarget sequences in the ARB database, further testing of the corresponding probe sequence can commence. It is important to note that the probe sequence is the *reverse complement* of the target string, as FISH probes target transcribed rRNA.
10. Confirm probe specificity against all publicly available DNA sequences using the BLASTN program at the National Center for Biotechnology Information website (*see* **Materials** and **Note 6**). An exact match will have a score (bits) in the BLAST Search Results descriptions output twice the value of the number of nucleotides of the submitted probe sequence, for example, an 18-mer probe will have an exact match score of 36. Sequences with mismatches to the probe sequence will have scores less than 36. Also, confirm the probe sequence is reverse complement by checking that "Strand = Plus / Minus" for a number of subject sequences in the BLAST Search Results alignments output. A target string will result in "Strand = Plus / Plus" (unless the subject sequence has been submitted to the databases as the reverse complement).
11. Confirm that the probe sequence has a melting temperature of ≥57°C using the nearest neighbor method (NN T_m, calculated using 50 m*M* NaCl and 50 μ*M* oligo). This can be performed on-line using websites such as the Biopolymer calculator or Primer calculator (*see* **Materials**). Empirical observations led us to believe that probes with NN T_m ≥57°C have a greater chance of success using the standard FISH protocol described below (originally described by Amann and coworkers *[7]*) likely due to the hybridization and wash temperatures employed (46° and 48°C, respectively). If the NN T_m of the probe is <57°C, the T_m can be raised by increasing the probe length, often a one or two base extension is sufficient. This requires revisiting the ARB database and ensuring any additional nucleotides added to the probe do not compromise probe specificity.
12. Self-complementarity of oligonucleotides (hairpins or dimers) can result in disruption of duplex formation between probe and target sequences. However, we have noted no correlation between probe hairpin or dimer formation potential (up to consecutive 4 bp) and success of the probe in FISH analysis.
13. Check probe accessibility. Ribosome accessibility to probes is a well-recognized limitation with the FISH method ***(1)***. Fuchs et al. ***(8)*** systematically evaluated the accessibility of the *Escherichia coli* ribosome to more than 200 oligonucleotides complementary to the entire length of the *E. coli* 16S rRNA and found regions of

high and low relative probe accessibility (*see* **Materials**). These data can be used as a rough guide to regions of the 16S rRNA molecule which should be avoided as target sites (<10% relative accessibility) if possible. However, organisms phylogenetically remote from *E. coli* may be expected to have different ribosomal higher order structure and therefore different probe accessibility profiles. In these instances the *E. coli* accessibility profile may be of limited value. If the FISH probe does not work, and poor accessibility is suspected, accessibility may be improved by the use of helper probes that are unlabeled oligonucleotides targeting adjacent regions to the FISH probe and in theory help open the target site ***(9)***.

14. Name the probe. A number of naming systems exists for 16S rRNA-directed oligonucleotide probes. The most common shorthand nomenclature in use is a three-letter abbreviation of the target group followed by the nucleotide position that the 3' end of the probe hybridizes to, usually according to standard *E. coli* numbering ***(10)***. For example, the commonly used FISH probe EUB338 ***(11)*** targets most *Bacteria* (Eubacteria) and the 3' binding position of the probe is 338 (the probe hybridizes to positions 338–355). Recently a more comprehensive naming system has been proposed ***(12)*** whereby several features of the probe are indicated in the name, including the target gene, target group, target group level (e.g., domain, division, genus), 3' end of probe, and probe length. For example, by this system EUB338 is named S-D-Bact-0338-a-A-18. Accession numbers also have been used to identify probes (*http://soul.mikro.biologie.tu-muenchen.de/ORS/*), similar to the system of unique accession numbers used to identify DNA sequences submitted to the public databases.

3.2 Probe Evaluation

Once designed, a probe can be synthesized and evaluated. This process usually involves hybridizing the probe to pure cultures of target organisms and nontarget organisms (with the fewest mismatches to the probe sequence) at a range of stringencies. Stringency can be adjusted via a number of parameters, such as temperature and formamide (denaturant) concentration. We routinely use the method described by Manz et al. ***(7)*** in which stringency is varied using formamide concentration at set hybridization and wash temperatures. The objective is to determine the range of stringencies (formamide concentrations) at which the probe specifically hybridizes to the target organisms but not to the nontarget organisms. The optimal stringency usually is taken as the highest formamide concentration before specific hybridization signal is lost.

1. Synthesize FISH probes. Fluorescently labeled oligonucleotide probes can be synthesized commercially for approx $60 (0.02 µmol synthesis scale) and provide enough probe for several thousand FISH reactions. Probes can be ordered on-line from companies such as Interactiva (Ulm, Germany) or Genset (worldwide) (*see* **Materials**). Fluorochromes are typically (and more cheaply) attached to the 5' end of the oligonucleotide and commonly available in fluorescein, Cy3

and Cy5. FISH probes can be synthesized and labeled in the laboratory, but this is no longer time or cost effective compared to commercially available probes.

2. Aliquot labeled probes into 2.5-μg aliquots in individual microcentrifuge tubes to avoid excessive freeze–thawing of probe stocks. Store stocks in the dark at –20°C. Resuspend each probe stock in a total volume of 50 μL of sterile milli-Q water to prepare a working concentration of 50 ng/μL (approx 9 pmol/μl for an 18-mer probe).
3. Sample fixation (*see* **Note 7**). For Gram-negative microorganisms add three volumes of 4% paraformaldehyde fixative to one volume of sample and hold at 4°C for 1–3 h. Pellet the cells by centrifugation (5000 *g*) and remove fixative. Wash the cells in 1× PBS and resuspend in 1× PBS to give $10^8 - 10^9$ cells/mL. Add one volume of ice-cold 100% ethanol and mix. Fixed cells can be spotted onto glass slides or stored at –20°C for several months. For Gram-positive microorganisms add one volume of 100% ethanol fixative to one volume of sample and hold at 4°C for 4–16 h. Pellet fixed cells by centrifugation (5000 *g*) and remove fixative. Wash the cells in 1× PBS and resuspend in 1× PBS to give $10^8 - 10^9$ cells/mL. Add one volume of ice-cold ethanol and mix. Ethanol-fixed cells should be prepared freshly for hybridization, as these samples do not store well.
4. Samples to include for probe evaluation are:
   - Fixed cells of log phase pure culture of a target organism (*see* **Note 8**)
   - Fixed cells of log phase pure culture of a nontarget organism (with fewest mismatches to the probe being evaluated).
5. Briefly vortex-mix fixed samples to resuspend settled material and apply 3–5 μL to wells on a Teflon-coated slide (*see* **Note 9**), air-dry thoroughly (to prevent cells detaching in subsequent steps), and dehydrate slides in an ethanol series (3 min each in 50%, 80%, and 98% ethanol). Slides can be stored at –20°C, but preferably should be hybridized soon after cell fixation and application.
6. For probe evaluation use a range of formamide concentrations in 10% increments, for example, 0–40% formamide. This will require five replicate slides, one for each formamide concentration. Determination of optimal formamide concentration can be refined by using smaller formamide concentration increments in the range of the broadly determined optimum.
7. Prewarm the hybridization oven to 46°C. Freshly prepare hybridization buffer in a 2-mL microcentrifuge tube (one tube of 2 mL of buffer per slide) in the following order:

   360 μL of 5 *M* NaCl (final concentration 0.9 *M*)

   40 μL of 1 *M* Tris-HCl (final concentration 20 m*M*, pH 7.2)

   x μL of 100% formamide (*see* **Table 1** and **Note 10**)

   y μL of autoclaved milli-Q water (according to volume of formamide; *see* **Table 1**)

   2 μL of 10% SDS (final concentration 0.01%; *see* **Note 10**)
8. Probe combinations for probe evaluation can include:
   - Test probe and domain-level probe (*see* **Note 11**)

Table 1
Formamide Volumes for Hybridization Buffer

% Formamide	Formamide volume *x* (µL)	Milli-Q water volume *y* (µL)
0	0	1598
5	100	1498
10	200	1398
15	300	1298
20	400	1198
25	500	1098
30	600	998
35	700	898
40	800	798
45	900	698
50	1000	598

- Nonsense probe (for nonspecific incorporation of probes into sample, *see* **Note 12**)
- No-probe control (for autofluorescence *see* **Note 13**)

9. Add 8 µL of hybridization buffer to each well containing sample on the slide. Fold a paper towel into a rectangle slightly larger than the slide, place the folded towel into a 50-mL polypropylene tube, and pour remaining hybridization buffer onto the paper towel. This prevents evaporation of buffer in the wells during hybridization. Add 0.5 µL of each probe (and competitor probe if required, *see* **Note 5**) at the working concentration of 50 ng/µL, and mix carefully with pipet tip (avoid touching surface of slide with pipet tip as this will disturb attached cells). Place the slide in the 50 mL tube containing the moistened towel. Screw on cap and place horizontally into hybridization oven at 46°C for 1–2 h.
10. During hybridization, prepare 50 mL of wash buffer in a fresh 50-mL polypropylene tube appropriate for hybridization buffer formamide concentration used (*see* **Table 2**) in the following order and prewarm to 48°C in a water bath:
    z µL of 5 *M* NaCl (*see* **Table 2**)
    1 mL of 1 *M* Tris-HCl (final concentration 20 m*M*, pH 7.2)
    Autoclaved milli-Q water up to 50 mL
    50 µL of 10% SDS (final concentration 0.01%)
11. Following hybridization, rinse wells immediately with 48°C wash buffer into the hybridization tube, using a pipet. Carefully remove slide from the hybridization tube, place into wash buffer tube, and hold at 48°C for 10–15 min. Remove slide from wash buffer, rinse briefly in a beaker of ice-cold distilled water, and thoroughly dry slide using compressed air (*see* **Note 14**). Rapid transfer of slides during these steps prevents cooling which can lead to nonspecific probe binding.
12. Mount slides in antifading solution such as Citifluor (**which is toxic**; avoid inhalation and contact with skin). Apply a thin film of Citifluor to the slide and place

Table 2
NaCl Concentrations of Wash Buffers According to Formamide Concentration in Hybridization Buffer

Percent formamide in hybridization buffer	5 *M* NaCl volume *z* (μL)	Final NaCl concentration of wash buffer (*M*)
0	9000	0.900
5	6300	0.630
10	4500	0.450
15	3180	0.318
20	2150	0.215
25	1490	0.149
30	1020	0.102
35	700	0.070
40	460	0.046
45	300	0.030
50	180	0.018

a large coverslip over the slide to cover all wells. Press coverslip down gently to remove excess Citifluor.

13. Observe slides using an epifluorescence or confocal laser scanning microscope, starting with the lowest formamide concentration and working upwards:
    - No-probe wells—observe in all available channels for autofluorescence of sample.
    - Nonsense probe wells—observe in appropriate channel for probe fluorescence, confirm no fluorescence occurs.
    - Test probe and domain-level probe wells—observe in appropriate channels, confirm positive fluorescence and note formamide concentration (*see* **Note 15**).

    The optimal stringency for the test probe usually is taken as the highest formamide concentration before specific hybridization signal is lost. The window of specific hybridization stringencies is between the lowest formamide concentration at which the nontarget organism shows no fluorescence and the optimal probe stringency. Often nontarget organisms do not fluoresce, even at 0% formamide.

4. Notes

1. It is not recommended to design probes based on a single sequence, as sequencing errors could be present that may be inadvertantly incorporated into the probe. Multiple sequences reduce the chances of sequencing errors affecting probe design because the identified region must be identical in all target sequences. Where possible use full-length sequences for probe design, as this provides the maximum possible sequence data to locate potential probe sites. Do not include

short sequences (<500 nt) in the design if possible as they may have little or no overlapping (comparable) regions on which the probe design is based.

2. The PT_SERVER searches for patterns (such as regions specific to target sequences) in special searchable database files, which are essentially fragmented versions of standard ARB database files. Pattern searches cannot be performed directly on standard database files, hence the need for defining the PT_SERVER. Before the probe design tool can be used, searchable database files must be created from a standard database file as follows:
   a. Open PT-SERVER administration window: *E*tc/Probe *F*unctions/PT_SERVER *A*dmin…
   b. Select a PT_SERVER template into which your database file will be loaded, for example, SSU_rRNA.arb would be appropriate for a 16S rRNA gene database.
   c. Click on UPDATE SERVER under Functions. Updating takes several minutes, and will overwrite any preexisting files in the selected template. Therefore, caution should be exercised if several people are using the one ARB program, as the potential exists for multiple users to update the same template.
3. Probes are usually designed to target a monophyletic group of sequences, and such a group may comprise short sequences that should be excluded in the design process:
   a. Mark target group of sequences.
   b. Open Search and Query window: Species/Search and Query.
   c. Search species that are marked: under DATABASE SEARCH click on "Search species that" in left-hand options and "are marked" in right hand options. Click on SEARCH. Marked sequences should appear in HITLIST window.
   d. Keep species that match the query: Under DATABASE SEARCH click on "Keep species that" in left-hand options and "match the query" in right-hand options.
   e. Under QUERY highlight the nucleotide (nuc) Search field and type >500 in Search string.
   f. Click on SEARCH. Sequences less than 500 nt long will be removed from the HITLIST window, but will remain marked. To unmark these short sequences click on MARK LISTED UNMARK REST. Only sequences greater than 500 nt long in the target group will be marked in the database.

   However, short sequences removed from the target group may contain probe sites inferred from analysis of full-length sequences, and this can be taken into account using the maximum number of nongroup hits. For example, if the monophyletic target group contains ten sequences, four of which are too short to include in the design process, then four should be entered into the Max. non group hits field, as these short sequences may contain target group probe sites inferred from the six full-length sequences. If Max non group hits is left at the default setting of zero, then the program may discount a potentially useful target group probe site because a short target sequence containing that site is seen as a nontarget sequence.

4. If no potential probe sites are found for a group of target sequences the message "There are no results" will be displayed in the PD RESULT window. Less than optimal probe sites may be found for the target group by reducing the "stringency" of the design parameters. For instance, reducing the Min group hits (%) or increasing Max. non group hits may result in the location of potential probe sites. These type of suboptimal probes will usually either not target all sequences in the target group, or target sequences outside the target group. However, depending on the application, suboptimal probes may be adequate as FISH probes, usually in combination with other probes.
5. Nonspecific hybridization of a FISH probe to nontarget sequences with a single mismatch to that probe may not be discriminated from specific target sequence hybridization regardless of stringency. Inclusion of an unlabeled oligonucleotide complementary to these nontarget sequences (competitor probe) should prevent hybridization of the FISH probe (and therefore nonspecific signal) by competing for the target site. Competitor probes are often denoted by a lowercase c preceding the name of the FISH probe they are competing with, for example, cPLA886 is the competitor probe for planctomycete-specific FISH probe PLA886 ***(13)***.
6. A BLAST search also serves to confirm that the probe sequence has been transcribed correctly and is in the correct orientation. Where possible probe sequences should be cut and pasted between programs rather than typed manually.
7. Gram-positive cells over crosslink with paraformaldehyde (PFA) which can result in reduced permeability to oligonucleotide probes, hence ethanol has been suggested as an alternative fixative ***(14)***. In addition, enzyme pretreatments can aid in permeabilization of Gram-positive cell walls, such as lysozyme ***(15)***, mutanolysin ***(16)***, and numerous other chemical pretreatments ***(17)***. For fixation of microbial community samples containing a mixture of Gram-negative and-positive organisms, we recommend using the standard PFA fixation procedure as the samples can be stored successfully for longer periods and in most cases, the majority of cells will be sufficiently permeable for probe entry and hybridization (as determined by comparison of universal FISH probes to nucleic acid-binding dyes such as 4,6-diamidino-2-phenylindole [DAPI]).
8. It is a good idea to sequence the 16S rDNA of the target and nontarget pure cultures in the probe target zone to confirm the identity of the cultures and probe specificity. If no pure cultures of a target organism exist, a habitat sample known to contain the target organism should be used as the positive control. This is usually the case if a probe has been designed to target an environmental sequence (a sequence obtained using a culture-independent PCR-clone library approach).
9. Teflon-coated glass slides often are pretreated for use in FISH by cleaning in 10% KOH solution or warm detergent and coated with gelatin or silane. However, we have found commercially prepared slides can be used successfully for FISH without any pretreatment.
10. Store formamide in 2-mL aliquots at –20°C. After thawing an aliquot, restore at

4°C and use within a week; fresh formamide should be colorless. Formamide is toxic, so gloves should be worn when handling it and hybridization tubes should be tightly capped and incinerated after use. Add SDS last to avoid precipitation with the concentrated NaCl.

11. If the test probe targets a bacterium, the bacterial domain-level probe EUB338 (5'–GCTGCCTCCCGTAGGAGT; target site *E. coli* no. 338-355) should be used in concert with the test probe labeled with a complementary fluorochrome (e.g., test probe-Cy3 + EUB338-fluorescein). Similarly, for a test probe targeting *Archaea*, ARC915 (5'–GTGCTCCCCCGCCAATTCCT; target site *E. coli* no. 915-934) can be used. These domain-level probes can be used at all formamide concentrations ***(18)*** and act as positive controls for most microorganisms to confirm the FISH procedure was successful. Some groups of bacteria have mismatches to domain-level probes, such as the *Verrucomicrobia* and *Planctomycetes*, for which variants of EUB338 exist ***(18)***. Domain-level probes should not be used in combination with the test probe is if their target sites overlap.
12. A nonsense probe such as nonEUB338 (reverse complement of EUB338), which has no known rRNA target, can be included in the probe evaluation to ensure nonspecific incorporation of the probe into the sample does not occur. Nonspecific probe incorporation is usually only a significant problem with some habitat samples.
13. Many compounds present in the environment autofluoresce, including some cellular components (e.g., photosynthetic pigments, cofactor F420, some proteins), which can obscure specific FISH. This is particularly evident in habitats such as soils, sediments, and aquatic samples. The wavelengths at which autofluorescence occurs can be sample specific (e.g., rumen samples have high autofluorescence under blue excitation/green emission). Therefore, no-probe controls should be included to detect autofluorescent cells at different wavelengths.
14. Ensure that all droplets of water are removed from the wells as the probe can dissociate and leave the cells due to osmotic pressure.
15. Optimal probe stringencies also can be determined quantitatively using image analysis software ***(18)***.

Acknowledgment

We would like to thank many colleagues from Lehrstuhl für Mikrobiologie, Technische Universität Müchen but especially Rudi Amann (Max Planck Institute for Marine Microbiology, Breman) for introducing LLB to FISH.

References

1. Amann, R. I., Ludwig, W., and Schleifer, K. H. (1995) Phylogenetic identification and *in situ* detection of individual microbial cells without cultivation. *Microbiol. Rev.* **59,** 143–169.
2. Woese, C. R. (1987) Bacterial evolution. *Microbiol. Rev.* **51,** 221–271.
3. Amann, R., Snaidr, J., Wagner, M., Ludwig, W., and Schleifer, K.-H. (1996) In

situ visualization of high genetic diversity in a natural microbial community. *J. Bacteriol.* **178,** 3496–3500.

4. Giovannoni, S. J., DeLong, E. F., Olsen, G. J., and Pace, N. R. (1988) Phylogenetic group-specific oligodeoxynucleotide probes for the identification of single microbial cells. *J. Bacteriol.* **170,** 720–726.
5. DeLong, E. F., Wickham, G. S., and Pace, N. R. (1989) Phylogenetic stains: ribosomal RNA-based probes for the identification of single cells. *Science* **243,** 1360–1363.
6. Stahl, D. A. and Amann, R. I. (1991) Nucleic acid techniques in bacterial systematics. In *Development and Application of Nucleic Acid Probes* (Stackebrandt, E. and Goodfellow, M., eds.), John Wiley & Sons, New York, New York, 205–248.
7. Manz, W., Amann, R., Ludwig, W., Wagner, M., and Schleifer, K.-H. (1992) Phylogenetic oligodeoxynucleotide probes for the major subclasses of *Proteobacteria*: problems and solutions. *Syst. Appl. Microbiol.* **15,** 593–600.
8. Fuchs, B. M., Wallner, G., Beisker, W., Schwippl, I., Ludwig, W. and Amann, R. (1998) Flow cytometric analysis of the in situ accessibility of *Escherichia coli* 16S rRNA for fluorescently labeled oligonucleotide probes. *Appl. Environ. Microbiol.* **64**, 4973-4982.
9. Fuchs, B. M., Glöckner, F. O., Wulf, J., and Amann, R. (2000) Unlabeled helper oligonucleotides increase the in situ accessibility to 16S rRNA of fluorescently labeled oligonucleotide probes. *Appl. Environ. Microbiol.* **66,** 3603–3607.
10. Brosius, J., Dull, T. J., Sleeter, D. D., and Noller, H. F. (1981) Gene organization and primary structure of a ribosomal RNA operon from *Escherichia coli. J. Mol. Biol.* **148,** 107–127.
11. Amann, R. I., Krumholz, L., and Stahl, D. A. (1990) Fluorescent-oligonucleotide probing of whole cells for determinative, phylogenetic, and environmental studies in microbiology. *J. Bacteriol.* **172,** 762–770.
12. Alm, E. W., Oerther, D. B., Larsen, N., Stahl, D. A., and Raskin, L. (1996) The oligonucleotide probe database. *Appl. Environ. Microbiol.* **62,** 3557–3559.
13. Neef, A., Amann, R., Schlesner, H., and Schleifer, K.-H. (1998) Monitoring a widespread bacterial group: *in situ* detection of planctomycetes with 16S rRNA-targeted probes. *Microbiology* **144,** 3257–3266.
14. Roller, C., Wagner, M., Amann, R., Ludwig, W., and Schleifer, K.-H. (1994) *In situ* probing of Gram-positive bacteria with high DNA G+C content by using 23S rRNA-targeted oligonucleotides. *Microbiology* **140,** 2849–2858.
15. Beimfohr, C., Krause, A., Amann, R., Ludwig, W., and Schleifer, K.-H. (1993) *In situ* identification of lactococci, enterococci and streptococci. *Syst. Appl. Microbiol.* **16,** 450–456.
16. Erhart, R., Bradford, D., Seviour, R. J., Amann, R. I., and Blackall, L. L. (1997) Development and use of fluorescent *in situ* hybridization probes for the detection and identification of "*Microthrix parvicella*" in activated sludge. *Syst. Appl. Microbiol.* **20,** 310–318.
17. Davenport, R. J., Curtis, T. P., Goodfellow, M., Stainsby, F. M., and Bingley, M.

(2000) Quantitative use of fluorescent in situ hybridization to examine relationships between mycolic acid-containing actinomycetes and foaming in activated sludge plants. *Appl. Environ. Microbiol.* **66,** 1158–1166.

18. Daims, H., Brühl, A., Amann, R., Schleifer, K.-H., and Wagner, M. (1999) The domain-specific probe EUB338 is insufficient for the detection of all *Bacteria:* development and evaluation of a more comprehensive probe set. *Syst. Appl. Microbiol.* **22,** 434–444.

7

Evaluation of Alterations in the Tumor Suppressor Genes *INK4A* and *INK4B* in Human Bladder Tumors

Irene Orlow and Carlos Cordon-Cardo

1. Introduction

The progression through the cell cycle is monitored by positive and negative regulators. One family of negative regulators has been reported to act as cyclin-dependent kinase inhibitors (*CKI*) ***(1–3)***; and these, in turn, have been subdivided into two groups on the basis of sequence homology. The first *CKI* family includes $p21^{Cip1}$ ***(4–6)***, $p27^{Kip1}$ ***(7–9)***, and $p57^{Kip2}$ ***(10,11)***. The other *CKI* subgroup includes four members: $p16^{INK4A/MTS1/CDKN2A}$ ***(12,13)***, $p15^{INK4B/MTS2/CDKN2B}$ ***(14)***, $p18^{INK4C}$ ***(15)***, and $p19^{INK4D}$ ***(16)***. The *INK4A* and *INK4B* genes map to the short arm of chromosome 9 (9p21), where they are found in tandem spanning a region of approx 80 kilobases (kb) (**Fig. 1**). The *INK4A* and the *INK4B* genes encode for the p16 and the p15 proteins, respectively ***(12–14)***. These protein products form binary complexes exclusively with Cdk4 and Cdk6, inhibiting their function and, by doing so, inhibiting pRB phosphorylation during G1. Additional complexity results from the presence of a second *INK4A* product, termed $p19^{ARF}$ or $p14^{ARF}$ in humans (ARF is the acronym for alternative reading frame) ***(17–20)*** (**Fig. 1**). The $p19^{ARF}$ blocks the *mdm2*-induced *p53* degradation and transactivational silencing ***(21,22)***. The *INK4A* is altered in many cell lines and primary tumors ***(23–26)***. Furthermore, germ line mutations of the *INK4A* gene are found on patients with familial melanoma and pancreatic adenocarcinoma ***(27–28)***; and targeted deletion of the *INK4A* in murine models is associated with the development of spontaneous tumors ***(29,30)***. The *INK4A* gene is localized in a chromosomal area found to be frequently altered in bladder transitional cell carcinomas ***(31)***. This, in addition to the functional and genetic evidence that supports an important role for *INK4A*

From: *Methods in Molecular Biology, vol. 179: Gene Probes: Principles and Protocols*
Edited by: M. Aquino de Muro and R. Rapley © Humana Press Inc., Totowa, NJ

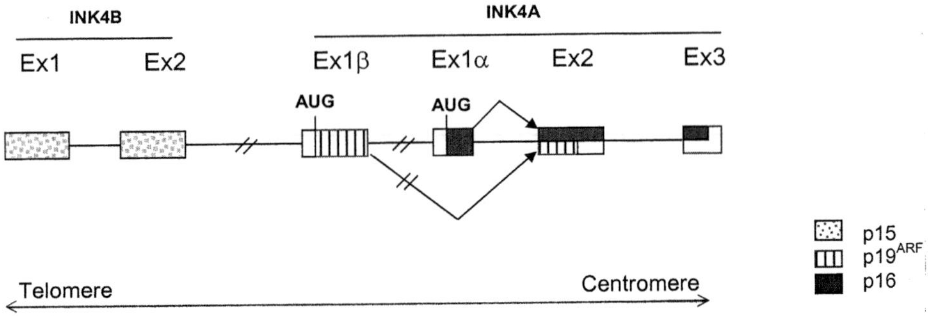

Fig. 1. Genomic organization of the *INK4A* and *INK4B* gene locus.

as a tumor suppressor gene in a variety of tumors prompted several groups to study the relevance of the *INK4A/B* locus in bladder tumors. Deletions, mainly homozygous or deletion of both alleles, of the *INK4A* and *INK4B* genes constitute a frequent finding in bladder tumors, and this loss is associated with early stages of the disease *(32)*. Patients bearing superficial tumors with *INK4A* homozygous deletions had a worse prognosis than those with wild-type *INK4A (33)*.

One of the most important and universal techniques developed to date to assess genomic organization is the immobilization of the nucleic acids onto a solid support membrane to analyze DNA sequence similarity by nucleic acid hybridization *(34)*. In this chapter, we present the optimal conditions used for the assessment of *INK4A* and *INK4B* gene deletions by Southern blot hybridization. In brief, the DNA samples are digested with a restriction enzyme, and the DNA fragments are resolved on an agarose gel. DNA denaturation and transfer to a nylon membrane follow this. The immobilized single-stranded DNA (ssDNA) is hybridized with a complementary isotopically labeled probe, specific to the target gene. The membrane is finally exposed to a sensitive screen or film, and the signal detected is proportional to the target DNA content. In some instances, the available DNA from scarce tumor samples is not sufficient for the analysis of gene deletions by Southern blot. Therefore, a semiquantitative method based on DNA amplification by the polymerase chain reaction (PCR) developed for the evaluation of alterations affecting the *INK4A* and the *INK4B* genes is discussed.

2. Materials

For the deletion analysis by Southern blot hybridization, the reagents and conditions described in the following subheadings are adapted for the use of the Probe Tech™ 1 System, which is an automated instrument that integrates electrophoresis and transfer of nucleic acids from a gel to a nylon membrane (Oncor, Gaithersburg, MD). The use of vacuum for the DNA transfer (*see* **Methods**) minimizes the gel handling and transfer time. This instrument is currently offered by Ventana (Tucson, AZ).

1. Genomic DNA: Tissue samples were embedded in optimal cutting temperature (OCT; Miles Laboratories, Elkhard, IN) cryopreservation compound, snap-frozen in isopentane precooled in liquid nitrogen, and stored at –70°C. Representative hematoxylin–eosin-stained sections of each frozen block were examined microscopically to confirm the presence of tumor, and only lesions with more than 50% neoplastic cells were included in the study (*see* **Note 1**). Normal tissues were obtained from all patients, either from a tumor-free area, such as skeletal muscle, or from peripheral blood. DNA was extracted by use of a nonorganic method (Oncor, Gaithersburg, MD) from paired normal tissue and tumor samples (*see* **Note 2**).
2. Buffer STE: 10 m*M* Tris-HCl, pH 7.5; 10 m*M* NaCl; 1 m*M* EDTA, pH 8.0.
3. DNA loading buffer: 0.42% Bromophenol blue, (10×) 0.42% xylene cyanol FF, 50% glycerol.
4. Buffer TE: 10 m*M* Tris-HCl, 1 m*M* EDTA, pH 8.0.
5. Buffer TAE (50×): 242 g of Tris base; 57.1 mL of acetic acid, 37.2 g of $Na_2EDTA{\cdot}2H_2O$. Complete to 1 L with distilled H_2O. For agarose run in the ProbeTech™ 1 instrument, it is necessary to use the TAE buffer provided by company.
6. TBE Buffer (20×): 108 g of 890 m*M* Tris base, 55 g of 890 m*M* boric acid, 40 mL of 0.5 *M* EDTA, pH 8.0 (20 m*M*). Complete to 1 L with distilled H_2O.
7. Salline sodium citrate (SSC) buffer (20×): 175.3 g of NaCl, 88.2 g of sodium citrate, pH 7.0. Complete to 1 L with distilled H_2O. Sterilize by autoclaving.
8. SDS 10%: 100 g of sodium dodicyl sulfate (SDS) in distilled water (final volume, 1 L); pH 7.2. Wear a mask while preparing this solution.
9. Prehybridization solution: Hybrisol I is a ready-to-use mix available from Oncor that contains 50% formamide, 6× SSC, 10% dextran sulfate, 1% SDS, sheared DNA, and modified Denhard's solution (cat. no. S4040).
10. Membranes: High-stringency positively charged nylon membranes. We used precut (10.9 cm × 13.5 cm) Sure Blot Nylon Hybridization Membrane provided by Oncor (Oncor, Gaithersburg, MD), but nylon membranes provided by other vendors could also be used.
11. Probes: A 0.5-kb complementary DNA (cDNA) fragment containing human *p16* sequences ***(13)*** and a 2-kb cDNA fragment containing human *p15* sequences ***(14)*** were used as probes to assess deletion and rearrangement of the *INK4A* and *INK4B* genes, respectively. A cDNA fragment containing glyceraldehyde phosphate dehydrogenase (GAPDH) sequences was used as control (American Type Culture Collection, ATCC no. 81141). We have also compared the signal of the

INK4A and *INK4B* genes to that obtained after hybridization with the pEFD126.3 probe (American Type Culture Collection, ATCC no. 57624).

12. Labeling Kit: The method of random primed DNA labeling is the method of choice to obtain DNA labeled to high activities. We obtained good results utilizing the kit offered by Boehringer Mannheim (cat. no. 1004760).
13 Columns: For the removal of unincorporated nucleotides from the labeled DNA probe, we utilized the Quick Spin columns provided by Boehringer Mannheim (cat. no. 100408) following the manufacturer's instructions. These are ready-to-use disposable G-50 Sephadex columns that can otherwise be packed manually into syringe barrels, although this is a tedious and time consuming step.
14. Gels: DNA samples digested with restriction enzymes were separated on 0.7–0.8% agarose gels. Many vendors offer agarose for routine use (gelling temperature approx 36°C), and they were all satisfactory. As an example, the agarose provided by Sigma (cat. no. A-9539) gives very good results. For the multiplex PCR we used a 7–9% polyacrylamide gel, prepared with 3.5 mL of acrylamyde/*bis*-acrylamide (29:1), buffer TBE to a final concentration of 1¥, 400 µL of 10% ammonium persulfate, 15 µL of TEMED, and distilled H_2O to a final volume of 20 mL. The acrylamide can be obtained from several vendors. We used the one provided by Fisher, as it has rendered excellent results.
15. Isotopes: (I) [a-^{32}P]dCTP (Dupont NEN Research Products, Boston, MA). The isotope should be stored at –20°C and should be kept behind acrylic shields at all times. [a-^{32}P]dCTP can be used for up to 2 wk. Special care needs to be practiced during probe labeling, purification, hybridization, and washes. (II) [a-^{33}P]dCTP (Amersham Life Science, Arlington Heights, IL). This isotope is an excellent alternative for the multiplex PCR assay. The signal is strong enough when used in PCR reactions run in thin polyacrylamide gels, and is safer to handle. Dupont-NEN also offers a stable reagent, which includes a dye (Easytides, cat. no. NEG 613H), which allows for a better visualization and stabilizes the reagent. This reagent can be stored at 4°C and can be used for up to 3 wk. Any radioactive disposable material (pipet tips, washing buffer, etc.) should be disposed according to the Institutional guidelines.
16. Restriction Enzyme: *TaqI* is available from many companies. We recommend using an enzyme with a concentration of approx 10 U per microliter (U/µL) to keep the volume to a minimum amount. As an example, the *TaqI* from Boehringer Mannheim (cat. no. 567671) gave us good and reproducible results.
17. Molecular Weight Markers: To verify the size or position of the DNA fragments and bands, molecular weight markers provided by any vendor can be used. Because of their easy interpretation we chose the following: 100-basepair (bp) ladder; 1-kb ladder; and l-*Hin*dIII Gibco, BRL).
18. Primers:
    (a) Controls. *GADPH* and *ANDRR* gene fragments that served as internal controls were amplified with the following sets of primers: GADPH—5' TGG TAT CGT GGA AGG ACT CAT GAC 3' (F) and 5' ATG CCA GTG AGC TTC CCG TTC AGC 3' (R) (fragment: 189 bp)(the HUMG3PDB sequence

corresponds to the glyceraldehyde gene and was obtained from Genbank); ANDRR—5' GTG CGC GAA GTG ATC CAG AA 3' (F) and 5' TCT GGG ACG CAA CCT CTC TC 3' (fragment: 296 bp) (the HUMARB sequence for the androgen receptor gene was obtained from GenBank); D9S196—5' ACC ACA CTG CGG GAC TT—3'(F) and 5'—GGG ATT ACA CCT CAA AAC CA—3' (R) (fragment: 260 bp).

(b) INK4A and INK4B. For the amplification of the $p19^{ARF}$-encoding exon 1, the following sets of primers were used: exon 1β (fragment 1, 439 bp) ***(19)***—5' TCC CAG TCT GCA GTT AAG G 3' (F) and 5' GTC TAA GTC GTT GTA ACC CG 3' (R); exon 1b (fragment 2, 160 bp)—5' AAC ATG GTG CGC AGG TTC 3' (F) and 5' AGT AGC ATC AGC ACG AGG G 3' (R). For the amplification of *INK4A* (exon 2) and *INK4B* (exon 2) the following sets of primers were used: *INK4A* (fragment 2a, 204 bp)—5' AGC TTC CTT TCC GTC ATG C—3' (F) and 5' GCA GCA CCA CCA GCG TG 3' (R) ***(27)***; *INK4A* (fragment 2c, 189 bp)—5' TGG ACG TGC GCG ATG C 3' (F) and 5' GGA AGC TCT CAG GGT ACA AAT TC 3' (R) ***(27)***; *INK4B* (fragment 2a, 198 bp)—5' CCC GGC CGG CAT CTC CCA TA 3' (F) and 5' ACC ACC AGC GTG TCC AGG AA 3' (R) ***(35)***.

3. Methods

3.1. Detection of INK4A and INK4B Deletions by Southern Blotting

3.1.1. Southern Blot

1. Digest the extracted DNA (7.5-µg aliquots) with the restriction endonuclease *TaqI* in a reaction mix containing 5 µL of spermidine, 5 µL of digestion buffer B (Boehringer Mannheim), 2 µL (20 U) of *TaqI*, and distilled H_2O to complete a final volume of 50 µL, at 65°C for 2–4 h (*see* **Note 3**).
2. Add loading buffer to the digested DNA and subject the mix to electrophoresis in 0.7% agarose gels/TAE buffer 1×, for 16 h at 18 V or 2 h at 80 V (*see* **Note 4** and **Fig. 2**).
3. Depurinate the DNA for 10 min with 0.25 *M* HCl, and neutralize for 30 min with 1 *M* NaOH.
4. Transfer the DNA onto nylon membranes, using 20 m*M* NaOH and 1 *M* ammonium acetate, and applying vacuum (5 ATM), for 90 min.
5. Fix the blotted DNA in a crosslinker oven (UV Crosslinker 2400, Stratagene) for 60 s, at 12,000 µJ/cm^2. Alternatively, the DNA can be fixed to the membrane by baking the membranes for an hour at 80°C. The membranes are air-dried and kept in a plastic bag until the next step.

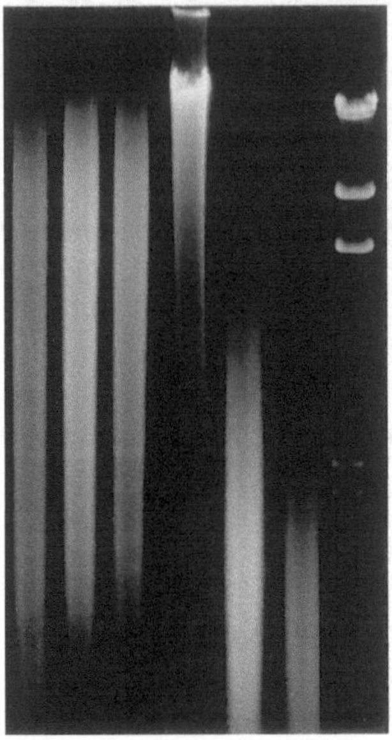

Fig. 2. DNA digestion with restriction endonuclease *Taq*I. Samples 1–3 depict completion of the digestion, while sample 4 shows an incomplete or partial digestion. Samples 5 and 6 depict partially degraded DNA samples. MW, molecular weight marker.

3.1.2. Probe Labeling

1. Label the DNA probes radioactively using the random oligonucleotide primed synthesis method as follows. Denature 50 ng of target probe in a total volume of 11 µL, by incubating at 95°C for 2 min, and quickly cool in ice water for 2 min.
2. Add 1 µL of dATP, dTTP, dGTP; 3 µL of distilled H_2O; 2 µL of hexanucleotides mix (Boehringer Mannheim); 5 µL of [α-^{32}P]dCTP (3000 Ci/mmol); and 1 µL of Klenow enzyme. Incubate the reaction tube for 30 min at 37°C.
3. Purify the labeled probe from the unincorporated nucleotides by chromatography through a Quick Spin™ G-50 Sephadex column (Boehringer Mannheim) following manufacturer's specifications.
4. Determine the specific activity of the labeled probe by measuring the radioactivity of a small aliquot (typically 1 µL) of sample, in a liquid scintilliation counter. The specific activity of the probes labeled by this method should be approx 1×10^7 cpm/membrane (*see* **Note 5**).

3.1.3. Hybridization

1. For newly prepared membranes, wash with 100 mL of 0.1 SSC–0.5% SDS, for an hour at 65°C.
2. Prehybridize membranes by placing them in a plastic bag or cylinder with 10 mL of Hybrisol I solution and incubate at 42–43°C for 1 h, with gentle shaking or rotation.
3. Denature the radiolabeled DNA probe by boiling or by incubating it at 95°C for 10 min, and immediately place it in the prehybridization solution for the hybridization of the DNA. Incubate overnight at 42–43°C (*see* **Notes 6** and **7**).
4. Wash the hybridized membranes at high stringency with 0.1× SSC–0.1% SDS at 70°C and expose to a sensitive film using intensifying screens for 24–72 h, at –70°C (*see* **Note 8**). If the membrane has been reprobed several times, it is possible that a longer exposure time will be needed (see step 3 below for reprobing procedure).
5. Quantify the band intensities by measuring the autoradiographic signals with the Ultrascan XL Laser Densitometer (Pharmacia LKB Biotechnology, Piscataway, NJ) ***(32)***; or by exposing the membranes to phosphoimage plates ***(33)***. The sensitized plates are then scanned by a phosphoimager (Bas 1000-Mac, Bio Imaging System Fujix, Fuji).
6. After documenting the band signal intensities, strip, prehybridize, and reprobe the membranes with the *INK4B* or the control-specific probe. For the probe-stripping, treat the membranes with 0.4 *N* NaOH and incubated at 45°C for 30–60 min. Drain the NaOH and neutralize the membranes with 0.2 *M* Tris-HCl, pH 7.5; 0.1× SSC, 0.1% SDS; at 45°C for 15 min. This is followed by a final wash in 0.1× SSC, 0.1% SDS at 52°C for 30–60 min.
7. Relative amounts of a given *CKI* gene (*INK4A* or *INK4B*) present in each sample are determined by comparing gene-specific hybridization signals with those obtained using the control probe, and are expressed as ratios (target-band signal)/(control-band signal). Some examples are shown schematically in **Fig. 3**. If the tumor tissues contained originally approx 70% of tumor cells, samples with <30% of the control signal are considered homozygously deleted; and those presenting 31–65% as hemizygously deleted for the *INK4A/B* specific genes.

3.1.4. Detection of INK4A and INK4B Gene Deletions in Bladder Tumors by Southern Blotting

Several independent groups of investigators showed that gene deletions involving the *INK4A* and *INK4B* locus are common events in bladder tumors. Applying the method described in the preceding, in a group of 110 primary bladder transitional cell carcinoma tumors, we found an overall frequency of deletions and rearrangements of 19% and 18%, respectively ***(32)***. Homozygous deletion (both alleles lost) of the *INK4A* and the *INK4B* genes was the most frequent finding; however, the exclusive loss of either gene was also detected in three tumors (**Fig. 4**). Hemizygous deletion (loss of heterozygosity,

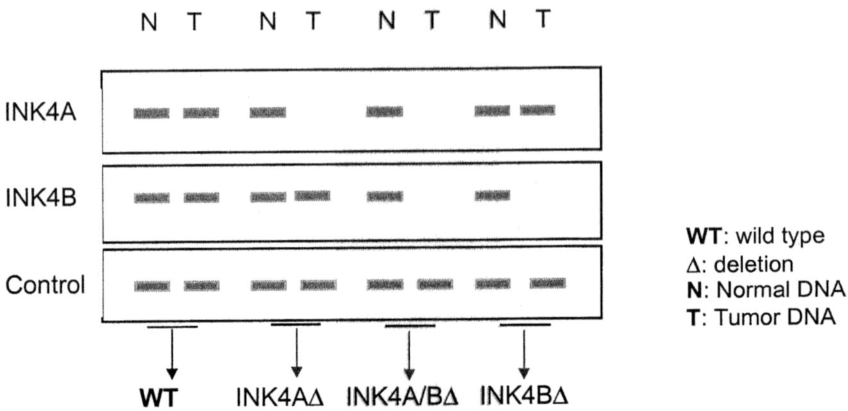

Fig. 3. Interpretation of *INK4A* and *INK4B* gene deletions by Southern blot analysis. The absence of *INK4A* and/or *INK4B* specific bands with the simultaneous presence of the internal control for DNA loading indicate specific gene or genes deletion as illustrated in the tumor samples. The wild-type case shows retention of all the specific bands.

or loss of one allele) of the *INK4A-p16* and/or *INK4B* genes was observed in eight tumors. Rearrangement of the two genes was indicated in three additional tumors (data not shown). Moreover, there was a statistically significant association between *INK4A-p16*/*INK4B* alterations and low stage/low-grade tumors ($p < 0.01$) *(32)*.

3.2 Detection of INK4A and INK4B Deletions by Comparative Multiplex PCR

In some instances, when the amount of DNA is limited, or when the genomic DNA is extracted from paraffin-embedded tissues, the evaluation of deletions by Southern blot is not possible. In these cases, and only when appropriate controls are available, the comparative multiplex PCR is a good alternative. In general, this method consists of the simultaneous amplification of genomic DNA using two sets of primers, one to the target gene sequence under study (*INK4A* or *INK4B* specific exons) and the other to an internal control gene sequence (e.g., the housekeeping gene *GAPDH*). Under optimal conditions, the absence of the target sequence in presence of the internal control indicates that the target gene fragment is deleted. In general, the preliminary experiments consist of the selection of internal controls, amount of genomic DNA, concentration of magnesium chloride ($MgCl_2$), dNTP mix, temperature, number of amplification cycles, and selection of cutoff points *(see* **Notes 9** and **10**).

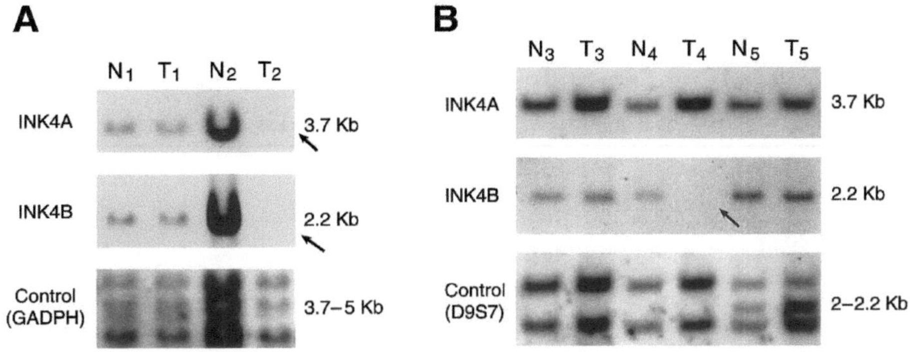

Fig. 4. DNA digestion of normal (N1, N2) DNA with *Taq* I followed by hybridization with specific *INK4A* and *INK4B* probes gives rise to specific bands of approx 3.7 kb and 2.2 kb, respectively. A third, nonspecific band of 1.0 kb (band not shown) was also noted. **(A)** Note the total absence (homozygous deletion) of the *INK4A* and *INK4B* specific bands in the tumor sample 2 (T2). The *GAPDH* probe served as internal control for DNA loading. **(B)** Exclusive deletion of the *INK4B* specific band. Here, D9S7 was used as reference for the DNA loading *(32)*.

Here, we describe in detail the conditions used for the detection of *INK4A-exon 1*β specific deletions. In addition, the cycling conditions for the evaluation of deletions of the *INK4A* and the *INK4B* gene will be listed (**Table 1**).

3.2.1 PCR Conditions

Each PCR reaction tube contained 50–100 ng of genomic DNA, 1× PCR buffer (Promega), 3.2 m*M* $MgCl_2$, 130 μM dNTP, 5% dimethyl sulfoxide (DMSO), 0.4 μ*M* of INK4A exon 1β primer, 0.4 μ*M* of each ANDRR primer, 0.5 U *Taq* polymerase (Promega), and 1 μCi of [α-^{33}P]dCTP. Samples are amplified as specified in the following. Cycling times correspond to reactions run in the Gene Amp PCR System 9700 (Perkin Elmer):

INK4A-ex1β and *ANDRR*:

95°C for 5 min; 2 cycles (95°C—15 s; 59°C—15 s; 72°C—30 s); 10 cycles (95°C—15 s, 55°C—15 s, 72°C—30 s); 20 cycles (95°C—15 s, 53°C—15 s, 72°C—30 s); 72°C—10 min

INK4A-ex1β and *GADPH*:

95°C—5 min; 2 cycles (95°C—15 s, 60°C—15 s, 72°C—30 s); 2 cycles (95°C—15 sc, 55°C—15 s, 72°C—30 s); 4 cycles (95°C—15 s, 54°C—15 s, 72°C—30 s); 19 cycles (95—15 s, 54°C—15 s, 72°C—30 s); 72°C, 10 min

Table 1
Comparative Multiplex PCR-Reaction Conditions

Amplified sets	Mg^{2+} (m*M*)	dNTP (m*M*)	Primers (each, pmol)	DMSO (%)	*Taq* pol (U)	[α-^{32}P]dCTP (mCi)	Final volume
INK4A-ex2a *ANDRR*	2.5	160	5	5	0.5	1	10 μL
INK4A-ex2c *D9S196*	2.5	160	5	5	0.5	1	10 μL
INK4A-ex1 *ANDRR*	3.2	130	4	5	0.5	1	10 μL
INK4A-ex1 *GADPH*	3.2	130	4	5	0.5	1	10 μL
INK4B-ex2a *ANDRR*	1.7	160	4	5	0.5	1	10 μL

INK4A-ex2a and *ANDRR*:

95°C—5 min; 25 cycles (95°C—30 s, 55°C—30 s, 72°C—30 s); 72°C, 10 min

INK4A-ex2c and D9S196:

95°C—5 min; 28 cycles (95°C—30 s, 55°C—30 s, 72°C—30 s); 72°C, 10 min

INK4B-ex2a and *ANDRR*:

95°C—5 min; five cycles (95°C—30 s, 63°C—30 s, 72°C—45 s); 5 cycles (95°C—30 s, 59°C—30 s, 72°C—45 s); 15 cycles (95°C—30 s, 55°C—30 s, 72°C—45 s); 72°C, 10 min

Additional reaction conditions are presented in **Table 1**.

3.2.2. Gel Electrophoresis and Image Analysis

1. To each tube containing 10 μL of reaction volume, add an equal volume of DNA loading buffer, and run an aliquot of 6–8 μL of this mix in nondenaturing 9% polyacrylamide gels at 40–45 W for 3–4 h.
2. Dry the gel and expose to a sensitive film. After obtaining the autoradiographic image, scan the signal by a phosphoimager (Bac 1000-Mac, Bio Imaging System Fujix, Fuji, Japan).
3. Express the presence of the *INK4A/B* specific fragments as the following ratio: (target-band signal)/(control-band signal).
4. All experiments need to be conducted at least twice, preferentially with duplicates (*see* **Note 11**).
5. Prepare a control curve, using tumor DNA samples known to be deleted for the *p16* and *p15* genes by a previous Southern blot analysis, as control DNAs ***(32,35)***.

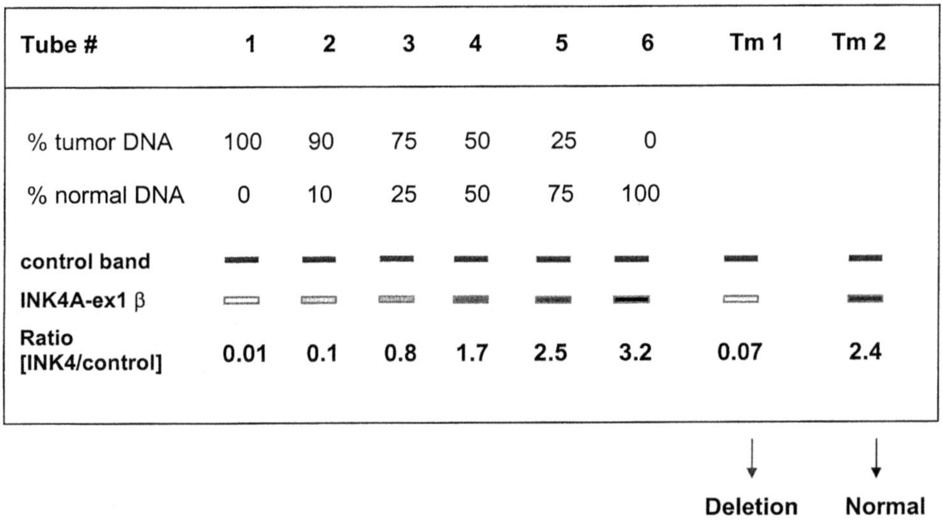

Fig. 5. Schematic representation of the *INK4A* gene deletion analysis by comparative multiplex PCR. Tubes 1–6 contain varying tumor-to-normal DNA mixtures that represent a range of the *INK4A*/*INK4B* gene content, varying from 0% of target (tumor sample control, known to be deleted for the *INK4A*/*B* genes) to 100% of target (normal DNA counterpart). These controls are amplified in parallel to the tumor samples to be analyzed, and the resulting (INK4/control) ratios are plotted in a control curve. The ratio (INK4/control) in tumor case 1 (Tm1) reveals that the sample contains <10% of normal DNA, and therefore is homozygously deleted. Tumor case 2 (Tm2) presents a normal *INK4* content.

This is necessary to validate the quantitative nature of the multiplex PCR method. Prepare varying mixtures of tumor DNA and normal genomic DNA, and amplify them as indicated previously for the samples to be analyzed in the study (**Fig. 5**). These tumor-to-normal DNA mixtures represent a range of the *INK4A*/*INK4B* gene content, varying from 0% of target (tumor sample control) to 100% of target (normal DNA counterpart). If the tumor sample contains approx 80% of tumor cells and 20% of normal cells, those samples presenting signals equal or <20% of the control signal will be considered homozygously deleted, and those presenting signals between 21% and 60% will be considered as heterozygously deleted for the studied gene fragment (**Fig. 5**; *see* **Notes 12** and **13**).

3.2.3. Detection of INK4A and INK4B Gene Deletions in Bladder Tumors by Multiplex PCR

Analysis of *INK4A* and *INK4B* deletions by Southern blot hybridization revealed that in bladder tumors there is a correlation between deletions of the *INK4A* gene and low-stage lesions ***(32)***. To confirm this finding we decided to

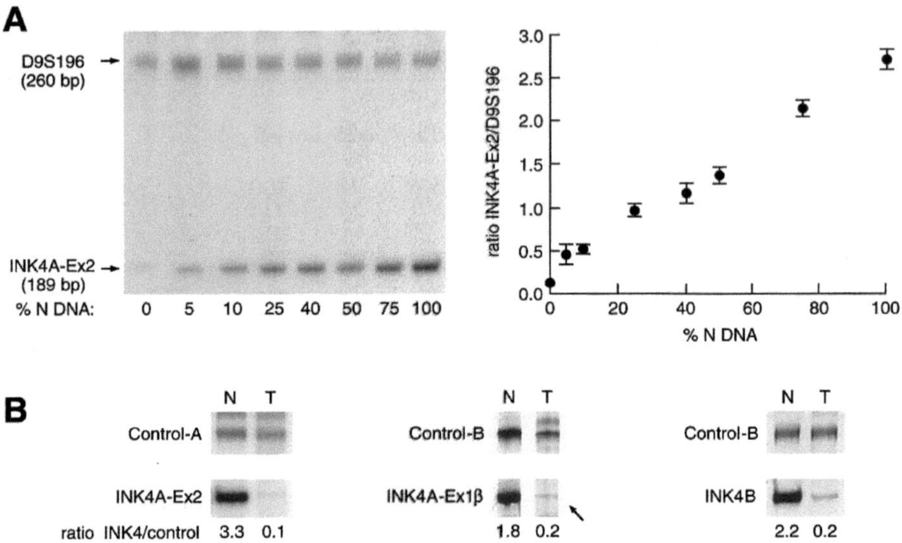

Fig. 6. (**A**) Standardization of the assay by incremental amplification of *INK4A*-ex2 with an increasing normal DNA target. The control curve was constructed with the relative (INK4A/control) ratios and the amount of normal DNA included in each sample. The vertical bars indicate the deviation of the duplicate values. (**B**) Proportional increment in the amount of *INK4A* (exons 2 and 1β) and *INK4B* (exon 2), expressed as the ratio (INK4A or INK4B/control). *GAPDH, ANDRR,* and *D9S196* were used as internal controls for DNA quality and loading. N, normal; T, tumor *(33)*.

analyze the status of the *INK4A/B* genes in a cohort of 121 patients diagnosed with superficial bladder tumors. For these tumors, the amount of DNA available was insufficient for Southern blot analysis; therefore, we applied the PCR based methodology described earlier. Using the comparative multiplex PCR, we found 17/121 (14.1%) cases with *INK4A* losses, including 13 homozygous (i.e., loss of both alleles) and 4 heterozygous (i.e., loss of one allele) deletions (**Fig. 6**). The comparative analysis of *INK4A* alterations and clinicopathological parameters revealed that homozygous deletions were the only mutations significantly associated with poor prognosis. Specifically, homozygous deletions of the *INK4A* gene were significantly associated with tumors displaying a large diameter (>3 cm) ($p = 0.003$) and the risk of recurrence among patients with *INK4A* deletions was 60% superior to that of patients without *INK4A* deletions ($RR = 1.58$, 95% *CI*: 0.77–3.26). This impact on the patient's outcome may be explained by the fact that the homozygous deletions of the *INK4A* completely inactivate two protein products (p16 and $p19^{ARF}$), which affect the two most critical tumor suppressor pathways controlling neoplasia, p16 through pRB and $p19^{ARF}$ through p53 ***(12,17–22)***.

4. Notes

1. For those cases in which there is a higher ratio of normal to tumor cells, a microdissection is necessary to assess allelic losses. This procedure will indirectly enrich the tissue sample with tumor cells.
2. It is extremely important to verify the purity and integrity of the genomic DNA. For this purpose, a small aliquot of genomic DNA (equal or smaller than 0.5 μg) can be run on an agarose gel. The size of the DNA should be of 10–20 kb for the nondigested sample.
3. It is advisable to monitor the completion of the DNA digestion with the restriction enzyme before adding the loading buffer. For this, we recommend analyzing a small aliquot (5 μL) of digested sample on an agarose gel. The digested sample should appear as a smear (**Fig. 2**, samples 1–3). If any of the samples is undigested (**Fig. 2**, sample 4), it is necessary to add another aliquot of restriction enzyme and to extend the incubation time one or two additional hours. Alternatively, the sample may contain an impurity that affects the enzymatic reaction. In this case, it is best to purify the DNA by precipitation with ethanol, for example, and restart the digestion reaction. If any sample appears degraded (**Fig. 2**, samples 5 and 6), the DNA should not be used for the Southern blot analysis; instead, another DNA extraction should be made for this purpose.
4. The electrophoresis can be monitored with an ultraviolet light-emitting lamp. The run should be stopped when the DNA fragments are well separated, and did not run out off the gel. For the analysis of the *INK4A* and *INK4B* genes, the electrophoretic run can be stopped when the xylene cyanol dye reaches about 3 cm from the bottom of the gel.
5. The random primed DNA labeling kit contains Klenow enzyme, which is temperature sensitive. For this reason, special care as with other enzymes is recommended, by using a benchtop cooler, or by aliquoting the enzyme and to store in two or three portions. For first time users, it is desirable to measure the radioactivity of the labeled probe. Each membrane should be hybridized with 50 ng of specific probe, and about 10^7 cpm/membrane. If the counts are low the following steps should be revised: (a) integrity of the DNA probe; (b) proper separation of the DNA strands before addition of enzyme and radiolabeled isotope; (c) incubation temperature; and (d) probe purification. Alternatively, Amersham Pharmacia Biotech offers DNA labeling beads (cat. no. 27-9240-01) that facilitate the storage of the reagents and do not require purification of the labeled probe.
6. It is important to place the heated probe immediately in the cylinder, without touching the membrane, as this can produce a dark dot or spot in the autoradiograph. We recommend adding 300–500 μL of Hybrisol I before heating the probe, to increase the volume. The *INK4A* and *INK4B* sequences are GC-rich and tend to reanneal quickly.

7. Membranes can be washed in a plastic resealable bag, tray, or glass cylinder. For the prehybridization and hybridization steps we placed the membranes in glass cylinders and the incubations were performed in a hybridization oven (Hybaid oven, Labnet). Alternatively, membranes can be placed in resealable plastic bags, and incubated in water baths.
8. The washes are monitored with a Geiger counter. We consider the background noise acceptable when the corners of the membranes show no more than 5–10 counts per second (cps).
9. The selection of appropriate internal controls is crucial for the optimization of the comparative multiplex PCR (mPCR). The control fragment corresponds to a sequence known to be normally present (nondeleted), and it has to amplify in PCR conditions that allow for the amplification of the target DNA sequence (*INK4A* or *INK4B* in this case). The control-specific primers cannot anneal to the target sequence.
10. The increase in the amount of amplicons stays exponential for a limited number of cycles, after which the amplification rate reaches a plateau (owing to substrate saturation of enzyme, as an example). In this latter phase, the quantitated amount of amplified product is no longer proportional to the starting amount of molecules. Therefore, it is necessary to identify a linear range of the reaction in which the quantitated amount of amplified target is proportional to the initial amount of target molecules, by establishing the optimal number of cycles for given amount of starting DNA material.
11. Ideally, the sets of primers should be selected so that the amplified target DNA is a little bit shorter than the amplified control. In general, fragments between 150 and 300 bp produce reliable results and allow the use of paraffin-embedded tissue as a source of genomic DNA (*see* **Note 14**).
12. In ideal conditions, when the tumor is homogeneous and lacks normal contaminating cells, the positive control sample (for homozygous or biallelic deletions) should give signals close to 0% of the signal detected in the normal, nondeleted control. However, many tumors, including bladder transitional cell carcinomas, contain a certain number of interstitial normal cells, and the normal DNA from these cells is coamplified in the PCR reaction. For example, if after examining the hematoxylin–eosin-stained tissues (by optical microscopy) the tumor case is defined as containing 20% of normal cells, then the cutoff would be as follows: 0–20% signal, homozygous deletion; 21–60%, heterozygous or partial deletion. The 60% cutoff is calculated as (20 [% normal cells] + 80/2 [% tumor cells, divided by 2 because the cells are losing only one gene copy] = 60%).
13. Because of the minute quantities of DNA used in the PCR-based methods, the multiplex PCR can be applied for the detection of gene deletions in DNA samples extracted from paraffin-embedded tissue. It is desirable, for these samples, to select primers that amplify fragments equal or smaller than 250 bp. If the fragment or fragments are larger, it is possible to notice absence of bands due to partial degradation of the DNA. This can be visualized as absence of the upper band (control) with presence of a lower (target) band, or as absence of both bands.

It is very important to select an internal control that will be slightly larger than the target fragment to avoid false-positives (absence of target due to degradation and not due to gene deletions).

Acknowledgments

This work was supported by the Public Health Service USA Grant CA-47538. The *p16* cDNA was kindly provided by Dr. Kamb and the *p15* cDNA by Dr. Beach. We thank Dr. Gloria Juan for her critical review of the manuscript.

References

1. Sherr, C. J. (1996) Cancer cell cycles. *Science* **274,** 1672–1677.
2. Clurman, B. E. and Roberts, J. M. (1995) Cell cycle and cancer. *J. Natl. Cancer Inst.* **87,** 1499–1501.
3. Cordón-Cardó, C. (1995) Mutation of cell cycle regulators. *Am. J. Pathol.* **147,** 545–559.
4. El-Deiry, W. S., Tokino, T., Velculescu, V. E., Levy, D. B., Parsons, R., Trent, J. M., et al. (1993) WAF1, a potential mediator of p53 tumor suppression. *Cell* **75,** 817–825.
5. Harper, J. W., Adami, G. R., Wei, N., Keyomarsi, K., and Elledge, S. J. (1993) The p21 cdk-interacting protein Cip1 is a potent inhibitor of G1 cyclin-dependent kinases. *Cell* **75,** 805–816.
6. Xiong, Y., Hannon, G. J., Zhang, H., Casso, D., Kobayashi, R., and Beach, D. (1993) p21 is a universal inhibitor of cyclin kinases. *Nature* **366,** 701–704.
7. Polyak, K., Kato, J. Y., Solomon, M. J., Sherr, C. J., Massague, J., Roberts, J. M., et al. (1994) p27Kip1, a cyclin-Cdk inhibitor, links transforming growth factor-beta and contact inhibition to cell cycle arrest. *Genes Dev.* **8,** 9–22.
8. Toyoshima, H. and Hunter, T. (1994) p27, a novel inhibitor of G1 cyclin-cdk protein kinase activity, is related to p21. *Cell* **78,** 67–74.
9. Ponce-Castañeda, V. M., Lee, M. H., Latres, E., Polyak, K., Lacombe, L., Montgomery, K., et al. (1995) p27 Kip1: chromosomal mapping to 12p12-12p13.1 and absence of mutations in human tumors. *Cancer Res.* **55,**1211–1214.
10. Lee, M.-H, Reynisdóttir, I., and Massague, J. (1995) Cloning of $p57^{KIP2}$, a cyclin-dependent kinase inhibitor with unique domain structure and tissue distribution. *Genes Dev.* **9,** 639–649.
11. Matsuoka, S., Edwards, M. C., Bai, C., Parker, S., Zhang, P., Baldini, A. (1995) $p57^{KIP2}$, a structurally distinct member of the $p21^{CIP1}$ Cdk inhibitory family, is a candidate tumor suppressor gene. *Genes Dev.* **9,** 650–662.
12. Serrano, M., Hannon, G. J., and Beach, D. (1993) A new regulatory motif in cell cycle control causing specific inhibition of cyclin D/CDK4. *Nature* **366,** 704–707.
13. Kamb, A., Gruis, N. A., Weaver Feldhaus, J., Liu, Q., Harshman, K., Tavtigian,

S. V., et al. (1994) A cell cycle regulator potentially involved in genesis of many tumor types. *Science* **264,** 436–440.
14. Hannon, G. J. and Beach, D. (1994) p15 ^{INK4B} is a potential effector of TGFβ-induced cell cycle arrest. *Nature* **371,** 257–261.
15. Guan, K. L., Jenkins, C. W., Li, Y., Nichols, M. A., Wu, X., and O'Keefe, C. L. (1994) Growth suppression by p18, a p16$^{INK4/MTS1}$ and p14$^{INK4B/MTS2}$-related CDK6 inhibitor, correlates with wild-type pRB function. *Genes Dev.* **8,** 2939–2952.
16. Okuda, T., Hirai, H., Valentine, V. A., Shurtleff, S. A., Kidd,V. J., and Lahti, J. M. (1995) Molecular cloning, expression pattern, and chromosomal localization of human CDKN2D/INK4D, an inhibitor of cyclin D-dependent kinases. *Genomics* **29**, 623–630.
17. Quelle, D. E., Zindy, F., Ashmun, R. A., and Sherr, C. J. (1995) Alternative reading frames of the INK4a tumor suppressor gene encode two unrelated proteins capable of inducing cell cycle arrest. *Cell* **83,** 993–1000.
18. Duro, D., Bernard, O., Della Valle, V., Berger, R., and Larsen, C. J. (1995) A new type of p16INK4/MTS1 gene transcript expressed in B-cell malignancies. *Oncogene* **11,** 21–29.
19. Mao, L., Merlo, A., Bedi, G., Shapiro, G. I., Edwards, C. D., and Rollins, B. J. (1995) A novel p16INK4A transcript. *Cancer Res.* **55,** 2995–2997.
20. Stone, S., Jiang, P., Dayananth, P., Tavtigian, S. V., Katcher, H., and Parry, D. (1995) Complex structure and regulation of the P16 (MTS1) locus. *Cancer Res.* **55**, 2988–2994.
21. Pomerantz, J., Schrieber-Agus, N., Liegoeis, N., Silverman, A., Alland, L., and Chin, L. (1998) The Ink4a tumor suppressor gene product, p19^{Arf}, interacts with MDM2 and neutralizes MDM2's inhibition of p53. *Cell* **92,** 713–723.
22. Zhang, Y., Xiong, Y., and Yarbrough, W. G. (1998) ARF promotes MDM2 degradation and stabilizes p53: ARF-INK4a locus deletion impairs both the Rb and p53 tumor suppression pathways. *Cell* **92,** 725–734.
23. Jen, J., Harper, J. W., Bigner, S. H., Bigner, D. D., Papadopoulos, N., Markowitz, S., et al. (1994) Deletion of p16 and p15 genes in brain tumors. *Cancer Res.* **54,** 6353–6358.
24. Spruck, C. H. III, Gonzalez-Zulueta, M., Shibata, A., Simoneau, A. R., Lin, M. F., Gonzales, F., et al. (1994) p16 gene in uncultured tumours. *Nature* **370,** 183–184.
25. Gruis, N. A., Weaver-Feldhaus, J., Liu, Q., Frye, C., Eeles, R., Orlow, I., et al. (1995) Genetic evidence in melanoma and bladder cancers that p16 and p53 function in separate pathways of tumor suppression. *Am. J. Pathol.* **146,** 1199–1206.
26. Reed, A. L., Califano, J., Cairns, P., Westra, W. H., Jones, R. M., Koch, W., et al. (1996) High frequency of p16 (CDKN2/MTS-1/INK4A) inactivation in head and neck squamous cell carcinoma. *Cancer Res.* **56,** 3630–3633.
27. Hussussian, C. J., Struewing, J. P., Goldstein, A. M., Higgins, P. A., Ally, D. S., Sheahan, M. D., Clark, W. H. Jr, et al. (1994) Germline p16 mutations in familial melanoma. *Nat. Genet.* **8,** 15–21.
28. Goldstein, A. M., Fraser, M. C., Struewing, J. P., Hussussian, C. J., Ranade, K.,

Zametkin, D. P., et al. (1995) Increased risk of pancreatic cancer in melanoma-prone kindreds with p16INK4 mutations. *N. Engl. J. Med.* **333,** 970–974.

29. Serrano, M., Lee, H.-W., Chin, L., Cordon-Cardo, C., Beach, D., and DePinho, R. (1996) Role of the INK4a locus in tumor suppression and cell mortality. *Cell* **85,** 27–37.
30. Kamijo, T., Zindy, F., Roussel, M. F., Quelle, D. E., Downing, J. R., Ashmun, R. A., et al. (1997) Tumor suppression at the mouse INK4a locus mediated by the alternative reading frame product $p19^{ARF}$. *Cell* **91,** 649–659.
31. Orlow, I., Lianes, P., Lacombe, L., Dalbagni, G., Reuter, V. E., and Cordon-Cardo, C. (1994) Chromosome 9 allelic losses and microsatellite alterations in human bladder tumors. *Cancer Res.* **54,** 2848–2851.
32. Orlow, I., Lacombe, L., Hannon, G. J., Serrano, M., Pellicer, I., Dalbagni, G., et al. (1995) Deletion of the p16 and p15 genes in human bladder tumors. *J. Natl. Cancer Inst.* **87,** 1524–1529.
33. Orlow, I., LaRue, H., Osman, I., Lacombe, L., Moore, L., Rabbani, F., et al. (1999) Deletions of the INK4A gene in superficial bladder tumors. Association with recurrence. *Am. J. Pathol.* **155,** 105–113.
34. Southern, E. M. (1975) Detection of specific sequences among DNA fragments separated by gel electrophoresis. *J. Mol. Biol.* **98,** 503–517.
35. Zariwala, M., Liu, E., and Xiong, Y. (1996) Mutational analysis of the p16 family cyclin-dependent kinase inhibitors $p15^{ink4b}$ and $p18^{ink4c}$ in tumor derived cell lines and primary tumors. *Oncogene* **12,** 451–455.

8

The β-Tubulin Gene Region as a Molecular Marker to Distinguish *Leishmania* Parasites

Alexis Mendoza-León, Luis Luis, and Clara Martínez

1. Introduction

Leishmania is a protozoan parasite belonging to the order Kinetoplastida, family Trypanosomatidae, and genus *Leishmania*. These parasites are the causative agents of the disease known as leishmaniasis, which in humans has a broad spectrum of clinical manifestations depending on the parasite species. The disease has a wide distribution throughout the world, being endemic in about 88 countries ***(1,2)***.

Until 1987, these parasites were grouped in complexes, which is a functional term used to show relationships among the species described, but has no taxonomic status. Lainson and Shaw (1987) have proposed two subgenera, *Leishmania (Viannia)* and *L. (Leishmania)*, based on the distribution of the parasites in the digestive tract of the sandfly vectors ***(3,4)***. The first subgenus is autochthonous of America, and includes those species related with cutaneous and mucocutaneous forms of the disease, for example, the *braziliensis* and *panamensis* species. The species of the *Leishmania* subgenus cause cutaneous and visceral leishmaniasis in both the New World and the Old World. Reference *Leishmania* strains of each subgenus have been defined by the WHO to facilitate parasite studies ***(4–6)***.

Different molecular markers of taxonomic value have been developed to differentiate the species of both subgenera. One of these molecular markers is the β-tubulin gene region, which has shown sufficient variability to discriminate between *Leishmania* subgenera ***(7–10)***.

From: *Methods in Molecular Biology, vol. 179: Gene Probes: Principles and Protocols*
Edited by: M. Aquino de Muro and R. Rapley © Humana Press Inc., Totowa, NJ

1.1. The Tubulin Genes in Leishmania sp.

The tubulin genes have a relatively high degree of conservation throughout evolution, at nucleotide levels also reflected in the polypeptide product. Many organisms present a multigene family encoding the tubulin proteins ***(11,12)***. In *Leishmania* the tubulin genes are repetitive, organized in tandem, with both genes, α—and β-tubulin, having unlinked repeats in the genome of the parasite, although dispersed β-tubulin genes have been observed in *L. (L.) major* ***(13–17)***.

Early experiments have established the polymorphism of the β-tubulin gene region of *Leishmania* and its value to distinguish among species of New World *Leishmania* ***(4,7,8)***. The polymorphism of the coding region of the β-tubulin gene and its flanking region also may be exploited for parasite identification with high sensitivity, specificity, and reliability ***(10)***. This chapter presents protocols for the preparation and characterization of the tubulin gene probes from *Leishmania* and its application as a molecular marker for *Leishmania* identification.

1.2. Restriction Fragment Length Polymorphism (RFLP) Analysis of the β–Tubulin Region in New World Leishmania

Initially, it was of cardinal importance to identify, isolate, and characterize nuclear multicopy sequences from the genome of *Leishmania* sp in particular, those sequences common to a number of species or those that are unique to single species.

Restriction enzyme analysis is an excellent tool to estimate genetic variation and rate of evolution at the nucleotide level, giving information about the relationship of two or more populations by simple comparison of the restriction patterns ***(18-21)***. The variation in fragment patterns is referred as restriction fragment length polymorphism (RFLP). These variations are related to loss or gain of cleavage sites for a particular endonuclease owing to random mutations such as base substitutions or insertion/deletion events.

We combine the use of restriction enzymes and hybridization with different DNA probes, to study comparatively the genome of New and Old World *Leishmania* (**Table 1**). The RFLP analysis of the *β*-tubulin gene region were used to estimate gene differences and establish relationships among the species of the parasite ***(8)***.

1.2.1. Restriction Enzymes Selection

The RFLP analysis requires appropriate selection of the restriction enzymes and hybridization probes. We selected endonucleases based on sequence recognition and cutting frequency. Moreover, the enzymes were selected with

Table 1
Designation of *Leishmania* Species and Other Organisms

Subgenus	Species[a]	Strain designation[b]	Abbreviation	Origin
Leishmania (L.)				
New World				
	mexicana	MHOM/BZ/82/BEL21	BEL21	Belize
		MNYC/BZ/62/M379	M379	Belize
		MHOM/VE/90/LCP9012	9012	Venezuela
	amazonensis	IFLA/BR/67/PH8	PH8C5[c]	Brazil
		MHOM/VE/72/AZV	AZV	Venezuela
		MHOM/VE/80/NR	NR	Venezuela
	garnhami	MHOM/VE/76/HM76	HM76	Venezuela
		MHOM/VE/76/JAP78	JAP78	Venezuela
Old World				
	donovani	MHOM/IN/80/DD8	DD8	India
	infantum	MHOM/FR/78/LEM75	LEM75	France
	major	MHOM/SU/59/P	P	USSR
	tropica	MHOM/SU/74/K27	K27	
Viannia (V.)				
	braziliensis	MHOM/BR/75/M2903	M2903	Brazil
		MHOM/BR/84/LTB300	LTB300	Brazil
		MHOM/PE/84/LC53	LC53	Peru
	peruviana	MHOM/PE/84/LC26	LC26	Peru
	guyanensis	MHOM/BR/75/M4147	M4147	Brazil
	panamensis	MHOM/PA/71/LS94	LS94	Panama
	naiffi	MDAS/BR/70/M5533	M5533	Brazil
	colombiensis	IHAR/CO/85/CL500[d]	CL500	Colombia
Other Kinetoplastida:				
	Trypanosoma brucei			EATRO427

(a) The nomenclature of *Leishmania* proposed by Lainson and Shaw (1987).
(b) The majority of these strains have been designated as reference strains by WHO.
(c) This is a cloned derivative of *L. (L.) amazonensis* stock PH8.
(d) Some evidences suggest the relationship of this species with *Endotrypanum* and its inclusion into the *Viannia* subgenus must be revised ***(23,24)***.

regard to the number of homologous fragments generated with the specific probe, which might be useful to distinguish among Old World and New World *Leishmania* at different levels. The most informative endonuclease to analyze the tubulin gene region are shown in **Table 2**. **Figure 1** shows restriction patterns to the endonucleases *Bam*HI and *Pst*I of the total genomic DNA of different species of the New World *Leishmania*, from the *Viannia* and *Leishmania*

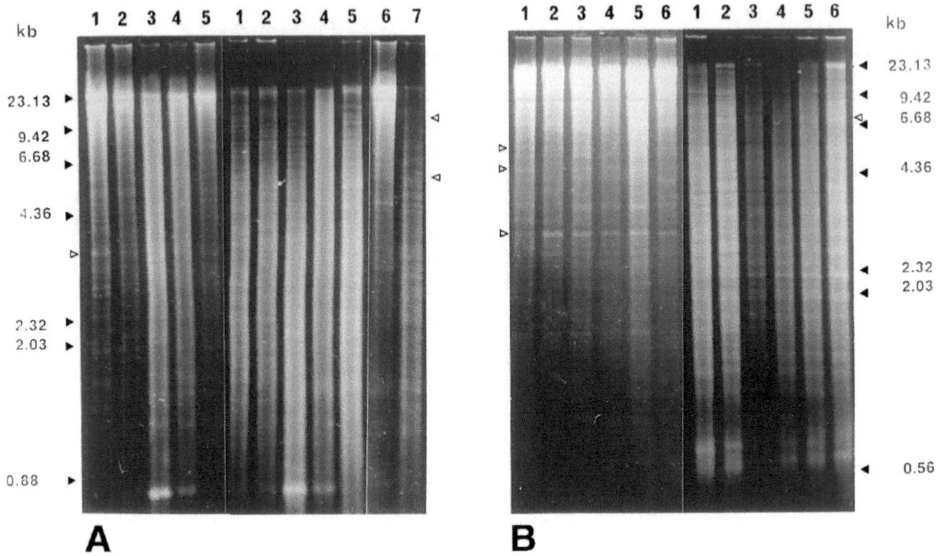

Fig. 1. Restriction fragment patterns of genomic DNA of New World *Leishmania*. Electrophoresis in a 1% agarose gel of genomic DNA of different strains of New World *Leishmania*, digested with *Bam*HI and *Pst*I. The *Viannia* subgenus (A) is represented by species *guyanensis* M4147 (lane *1*); *braziliensis* strain M2903 (lane *2*), LTB300 (lane *3*), LC53 (lane *4*); and *panamensis* LS94 (lane *5*). Genomic DNA from the species *infantum* strain LEM75, belonging to the *Leishmania* subgenus of the Old World *Leishmania*, digested with *Bam*HI (lane *6*) and *Pst*I (lane *7*), was included for comparison. The species of the *Leishmania* subgenus (B) include *mexicana* M379 (lane *1*); *amazonensis* strains PH8C5 (lane *2*), NR (lane *3*), HM76 (lane *4*), JAP78 (lane *5*), and AZV (lane *6*). The open arrowhead indicates the bands with the same molecular weight shared among the strains. The restriction fragments were fractionated at 30 V for 18 h and the gel stained with ethidium bromide. Molecular weight markers correspond to *Hin*dIII fragments of λ DNA.

subgenera. Given the complexity of *Leishmania* nuclear DNA, restriction pattern analysis showed characteristic fragments in all strains, with differences in fluorescence intensity suggesting the presence of repetitive sequences. We assume that many of these fragments are represented by minicircles release from the kinetoplast DNA (kDNA) network.

1.2.2. Probe Selection

The genomic clone of the β-tubulin region, pLgβ4, from New World *Leishmania* species *L. (V.) guyanensis* M4147 was isolated using a heterologous probe. The restriction map of the pLgβ4 was previously described ***(8)***. Two

Table 2
The Most Informative Restriction Endonucleases to Analyze the β-Tubulin Gene Region of *Leishmania* sp.

Enzyme	Sequence
*Bam*HI	5'-G GATC C-3'
*Eco*RII	5'-CC (A/T) GG-3'
*Hae*III/*Pal*I	5'- GG CC-3'
*Hin*dIII	5'- A AGCTT-3'
*Pst*I	5'- CTGCA G-3'
*Pvu*II	5'- C AG CTG-3'
*Sal*I	5'- G TCGAC-3'

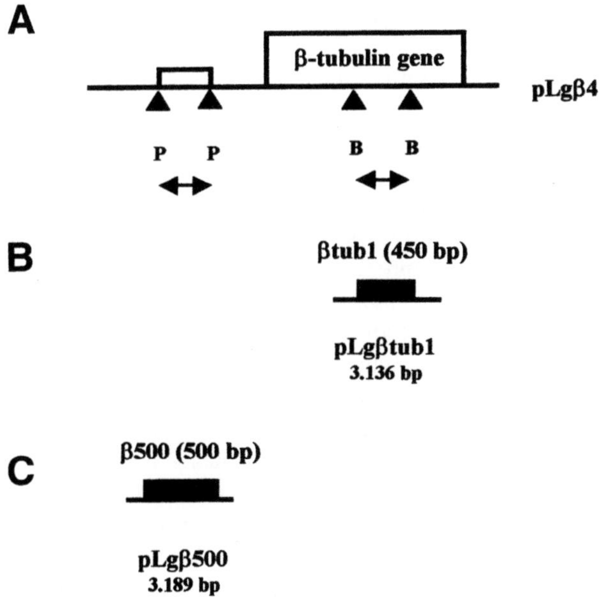

Fig. 2. The β-tubulin gene region of New World *Leishmania (Viannia) guyanensis*. The pLgβ4 clone and derived clones. The recombinant plasmid pLgβ4 **(A)** contains a 2.3-kb *Hin*dIII-genomic fragment that harbors the complete coding region of the β-tubulin gene from *L. (V.) guyanensis* M4147 strain, cloned in the pUC18 vector. The recombinant pLgβtub1 **(B)** corresponds to a *Bam*HI fragment of 0.45 kb that contains partial sequences of the coding region of the gene (the gene size corresponds to 1.32 kb). The pLgβ500 **(C)** contains a *Pst*I fragment of 0.5 kb from the upstream region outside the gene. Both are cloned in the pUC18 vector.

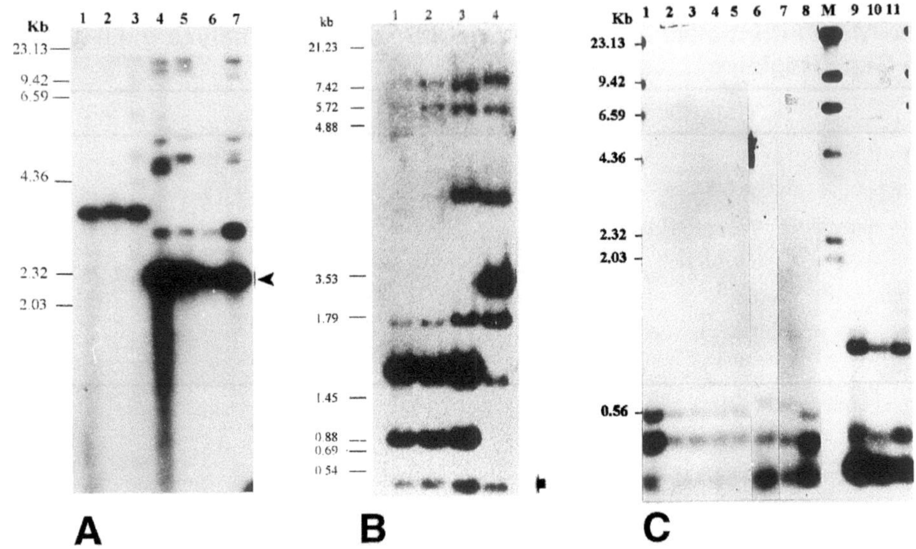

Fig. 3. The Differential Pattern of the β-Tubulin Gene Region of New World *Leishmania*. Total genomic DNA from different species of *Leishmania* representative of *Leishmania* and *Viannia* subgenera was digested with appropriate endonucleases and after electrophoresis in 1% agarose gel (described in **Fig. 1**) and blotting, the filter was hybridized with the proper probe. **(A)** Hybridization of *Hin*dIII genomic fragments with ^{32}P–pLgβ4; lanes *1* and *2*, *mexicana* strain BEL 21 and M379; lane *3*, *amazonensis* PH8C5; lanes *4* and *5*, *braziliensis* strain LTB300 and M2903; lane *6*, *guyanensis* M4147; and lane *7*, *panamensis* LS94. The arrow indicates the *Hin*dIII fragment cloned in pLgβ4. **(B)** Hybridization of *Pst*I genomic fragments from *Viannia* species with ^{32}P–pLgβ4; lanes *1* and *2*, *braziliensis* strain M2903 and recent fields isolate; *lane 3*, *guyanensis* M4147; and lane *4*, *panamensis* LS94. The arrow indicates the β500 fragment. **(C)** Hybridization of *Eco*RII genomic fragments with ^{32}P–pLgβtub1; lane *1*, *amazonensis* PH8C5; lanes *2* and *3*, *mexicana* BEL 21 and M379 strains; lane *4*, garnhami JAP78; lane *5*, *amazonensis* NR; lane *6*, *major* P; lane *7*, *tropica*; lane *8*, *mexicana* M9012; lane *9*, *braziliensis* M2903; lane *10*, *guyanensis* M4147; and lane *11*, *panamensis* LS94. Molecular weight markers correspond to *Hin*dIII fragments of λDNA and *Hae*III fragments of phage PM2 DNA.

different probes were constructed from the pLgβ4 (**Fig. 2**). The first one corresponds to pLgβtub1, which contains a *Bam*HI fragment of 0.45 kilobase (kb) from the coding region of the β-tubulin gene. This probe was used to evaluate the polymorphism of the coding region of the gene. A second probe, the pLgβ500, contain a *Pst*I-fragment of 0.50 kb located upstream of the coding region of the β-tubulin gene. The β500 was found in the intergenic region of the tandem arrays of the tubulin genes.

1.2.3. Typing New World Leishmania

An initial comparison of the β-tubulin gene region of different New World *Leishmania* strains was analyzed on a Southern blot of the genomic DNA digested with *Hin*dIII (**Fig. 3A**) and *Pst*I (**Fig. 3B**), hybridized with the pLgβ4 probe. The pLgβ4 probe detected differences among New World *Leishmania* species belonging to the *Leishmania* and *Viannia* subgenera. There were different *Hin*dIII or *Pst*I restriction patterns for the β-tubulin region among *mexicana* and *braziliensis* species. Multiple band patterns were observed in all *Leishmania* strains, indicating the presence of multiple copies of the β-tubulin gene in the *Leishmania* subgenera. An identical analysis was carried out with *Eco*RII (**Fig. 3C**) and hybridization with a pLgβtub1 probe. The results showed that there was sufficient polymorphism in the coding region of the β-tubulin gene among *Leishmania* parasites, distinctly at the subgenus level.

1.3. Differences between New World and Old World Leishmania

1.3.1. The β–Tubulin Region

An initial approach to differentiate between New World and Old World *Leishmania* was performed using a combination of restriction enzymes and hybridization with a heterologous tubulin probe. **Figure 4** shows a comparison of the *Pst*I pattern among species of *Leishmania* when a Southern analysis was carried out with the heterologous *Trypanosoma brucei* β/α-tubulin p1101 probe, which contains a complete unit of the β/α-tubulin gene isolated as a *Hin*dIII-fragment (**Fig. 4A**). The results showed clear differences among New World (lanes 1–3) and Old World (lanes 4–6) *Leishmania* strains belonging to the *Leishmania* subgenus, suggesting intrasubgenus variation. In contrast, for the same enzyme marked differences were found in the restriction patterns of species belonging to the *Viannia* subgenus (lanes 8–11) establishing an intergenera differentiation. The variation found in the species of *Leishmania* was sufficient to be distinguished from organisms of other genera of the Kinetoplastida order such as *T. brucei* (lane 7). Similar results were found when the homologous *Leishmania* β-tub1 probe was used (**Fig. 4B**). Again under this analysis the main differences among *Leishmania* parasites occurred at the subgenus level. Nevertheless, at present we are evaluating particular restriction patterns observed in some species, which may define particular species. This is the case of the *Pst*I pattern to *L. (V.) panamensis*, where an intragenic *Pst*I seems to be species specific ***(10)***.

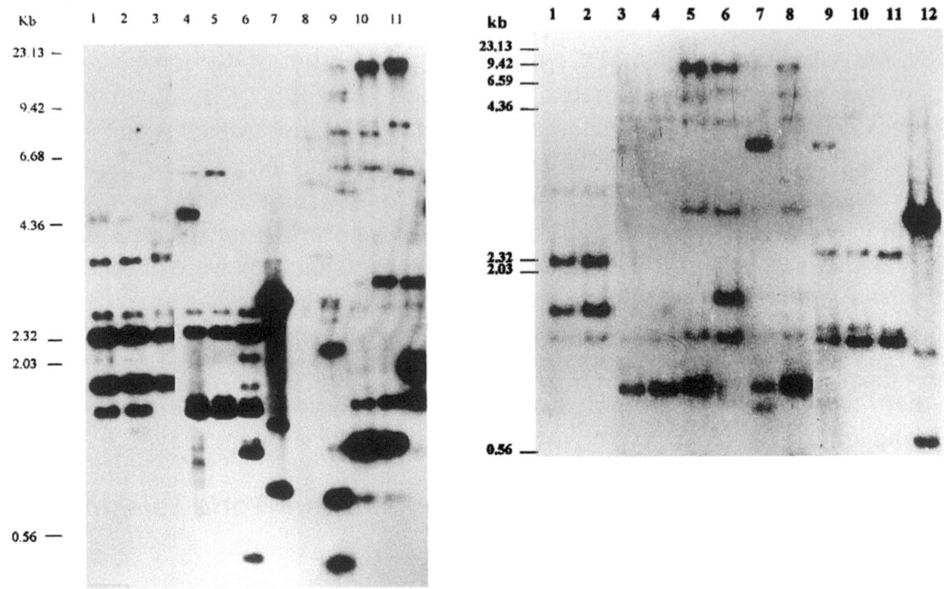

A **B**

Fig. 4. Differences between New World and Old World *Leishmania*. Total genomic DNA from different species of New World and Old World *Leishmania*, representative of *Leishmania* and *Viannia* subgenera, was digested with the endonuclease *Pst*I. After electrophoresis in 1% agarose gel and blotting, the filter was hybridized with the proper probe. **(A)** Hybridization with ^{32}P–*T. brucei* β/α-tubulin p1101 probe. The New World *Leishmania* belonging to the *Leishmania* subgenus include: lane *1*, *mexicana* BEL 21; lane *2*, *amazonensis* PH8C5; lane *3*; *amazonensis* NR. Old World *Leishmania* are represented by: lane *4*; *donovani* DD8; lane *5*, *infantum* LEM75: and lane *6*, *major* P. The *Viannia* subgenus species correspond to: lane *8*, *braziliensis* LTB300; lane *9*, *braziliensis* M2903; lane *10*, *guyanensis* M4147; and lane *11*, *panamensis* LS94. Lane *7*, *T. brucei*. **(B)** Hybridization with ^{32}P–pLgβtub1 probe. Lane *1*, *mexicana* BEL 21; lane *2*, *amazonensis* PH8C5; lane *3*, *braziliensis* LTB300; lane *4*, *braziliensis* M2903; lane *5*, *guyanensis* M4147; lane *6*, *panamensis* LS94; lane *7*, *peruviana* LC26; lane *8*, *braziliensis* LC53; lane *9*; *donovani* DD8; lane *10*, *infantum* LEM75: and lane *11*, *major* P. Lane *12*, *T. brucei*.

1.3.2. PCR-RFLP Analysis: New World vs Old World Leishmania

The applicability of the polymerase chain reaction (PCR), in combination with RFLP analysis of the PCR-amplified products, has been improved to analyze the variability of the β-tubulin gene region of *Leishmania sp.* Significant homologies of the β-tubulin gene of *L. (L.) amazonensis*, *L. (L.) major*, and *L.*

Table 3
Oligonucleotide Primers Designed to Analyze the β-Tubulin Gene Region of *Leishmania* sp.

Primer	Sequence	Specificity	PCR-product (bp)
Tub 1	5'-ATGCGTGAGATCGTTTCC-3'	All Leishmania	
Tub 6	5'-GGCGGCCTGCATCAT-3'	All Leishmania	
Tub 1/Tub 6	—		900
A2	5'-GACACGCGCTTGCGCACTCGT-3'	*Viannia*	
A10	5'-CCCCCTGCCTCGCCTGC-3'	*Viannia*	
A2/A10	—		375

(V.) guyanensis was demonstrated after comparison of nucleotides or amino acids sequences ***(21,22)***. We used this homology to generate oligonucleotide PCR-primers from conserved regions, and selected specific restriction enzymes to improve the PCR–RFLP analysis to study the variability of the gene in *Leishmania* genus. To conduct such analysis the following primers were designed from *L. (V.) guyanensis* sequence: *tub1* from nucleotide position 1 to 18 and *tub6* from nucleotide position 901–916 (**Table 3**). The PCR reaction generated an amplification product of 900 bp. Endonuclease such as *Ava*I, *Eco*RII, *Pst*I, and *Pvu*II were suitable to evaluate the variability of the amplified fragment. Example of this analysis is shown in **Fig. 5**, where differences between *Leishmania* and *Viannia* subgenera were found for *Pst*I and *Pvu*II endonuclease. The main differences found with this kind of analysis are at the subgenus level.

Mutation detection techniques such as denaturing gradient gel electrophoresis (DGGE) or single-strand conformation polymorphism coupled to the PCR procedure may provide an alternative approach for the analysis of nucleotide variations of the β-tubulin gene.

1.4. The Upstream Region of the β-Tubulin Gene of Leishmania (Viannia) Subgenus

Although a very similar restriction pattern could be observed when a heterologous probe is used, for example, *T. brucei* p1101, the homologous one, such as pLgβ4, may reveal other interesting sequences. This was the case for the β500 DNA sequence (for comparison see **Fig. 3B** and lanes 8–11 of **Fig. 4A**).

Subsequent analysis of the upstream region of the pLgβ4 clone demonstrated the presence of a 0.50-kb *Pst*I-fragment (β500 DNA). This fragment was common to species of the *Viannia* subgenus (**Fig. 3B**). Furthermore, this sequence

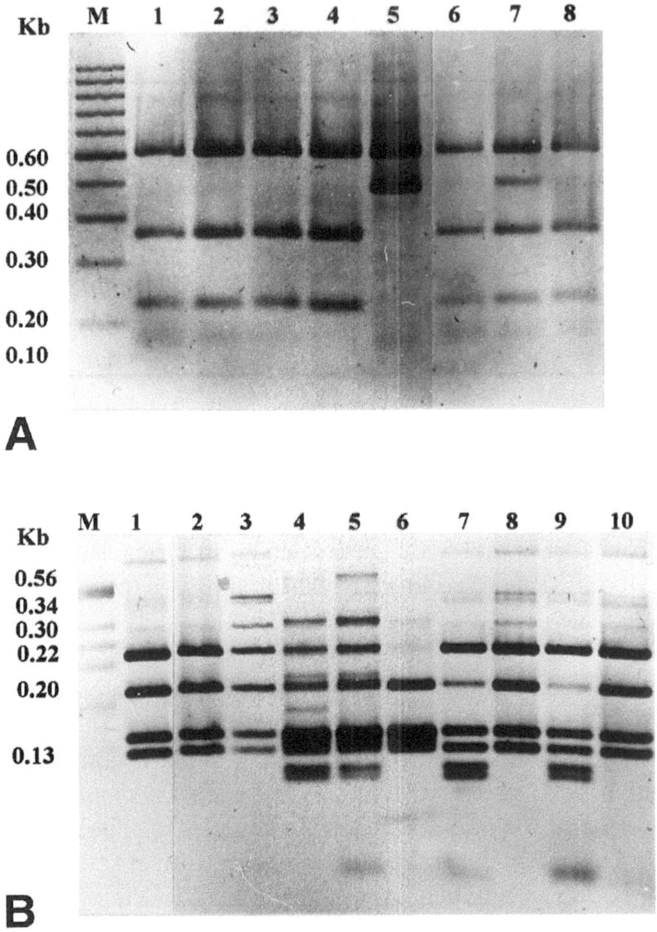

Fig. 5. PCR–RFLP Analysis of the Coding Region of the β-Tubulin Gene of *Leishmania* sp. This analysis can be improved by the use of specific primers designed from the conserved region of the coding sequence of the gene (*see* text). The primer set *tub1/tub6* generates a PCR product of 0.90 kb in all *Leishmania* tested (*see* **Fig. 6**). The RFLP of this fragment provides a discriminatory pattern at subgenus level among *Leishmania* strains. **(A)** *Pst*I digestion. Lane *1*, *mexicana* BEL 21; lane *2*, *garnhami* JAP78; lane *3*, *amazonensis* PH8C5; lane *4*, *mexicana* M379; lane *5*, *guyanensis* M4147; lane *6*, *donovani* DD8; lane *7*, *major* P; lane *8*, *tropica*. M, 0.10 kb DNA ladder. **(B)** *Pvu*II digestión. Lanes *1* and *2*, *garnhami* HM76 and JAP78; lane *3*, *mexicana* BEL 21; lane *4*, *braziliensis* M2903; lane *5*, *guyanensis* M4147; lane *6*, *naiffi* M5533; lane *7*, *tropica*; lane *8*; *donovani* DD8; lane *9*, *major* P; lane *10*, *mexicana* M9012. M, 1.0-kb DNA ladder.

was subcloned from pLgβ4 and the resulting clone, pLgβ500, was used to demonstrate the specificity of β500 DNA sequences for *Leishmania* (*Viannia*) subgenus. The results confirmed the presence of the β500 DNA sequence in reference strains of the *Viannia* subgenus such as *L. (V.) panamensis* LS94 (lane 4) and *L. (V.) braziliensis* M2903 (lane 5) (**Fig. 6**B). In addition, the demonstration of this sequence in field isolates of *Leishmania* was evaluated with a positive result (**Fig. 6B**, lanes 1–3). Surprisingly, the signal for this probe was absent in the *L. (V.) colombiensis* reference strain CL500 (lane 7). This species has been included in the *Viannia* subgenus, but other authors have proposed that its taxonomic status must be reviewed ***(23,24)***. In contrast, there was no sequence homology in the reference strains of the *Leishmania* subgenus, for example, *L. (L.) mexicana* BEL 21 (lane 6).

The specificity of this sequence was exploited further to develop a PCR assay, where only DNA derived from species of the *Viannia* subgenus could serve as templates for the β500 DNA amplification. The primer selection (A_2/A_{10}, **Table 3**) and amplification parameters were optimized to produce a highly sensitive assay able to detect a single parasite (Mendoza-León, *unpublished results*). A PCR assay using genomic DNA from representative species of both subgenera of *Leishmania* parasites and skin lesion samples from leishmaniasis patients confirmed the specificity of the β500 DNA sequence (**Fig. 6C**). As an internal control for the PCR assay, the internal region of the β-tubulin gene (described in **Subheading 3B-II**) was also amplified. This control evaluates possible false-negatives due to inhibition of the PCR reaction. We have isolated the β500 DNA sequences of other representative strains of *Leishmania* species from the *Viannia* subgenus and their sequence analyses are in progress to evaluate the variability of the sequence in this subgenus. Thus, we have defined this sequence as a molecular marker for the *Viannia* subgenus ***(24)***.

1.5. Conclusion and Perspectives

Our results have shown that the *Leishmania* β-tubulin genes show sufficient polymorphism to establish differences at the subgenus level. Thus, when the coding region is used as a molecular marker, *Leishmania* parasites can be distinguished at the subgenus level, although some species such as *L. (V.) panamensis* showed polymorphic sites in the coding region of the gene. The evaluation of this site as a marker of this species is in progress. The differences found in the coding region of these genes between *Leishmania* and other Kinetoplastida organisms suggest that a combination of PCR procedures with mutation detection techniques may be useful in developing a precise method to establish distinctions among *Leishmania* sp. at species level.

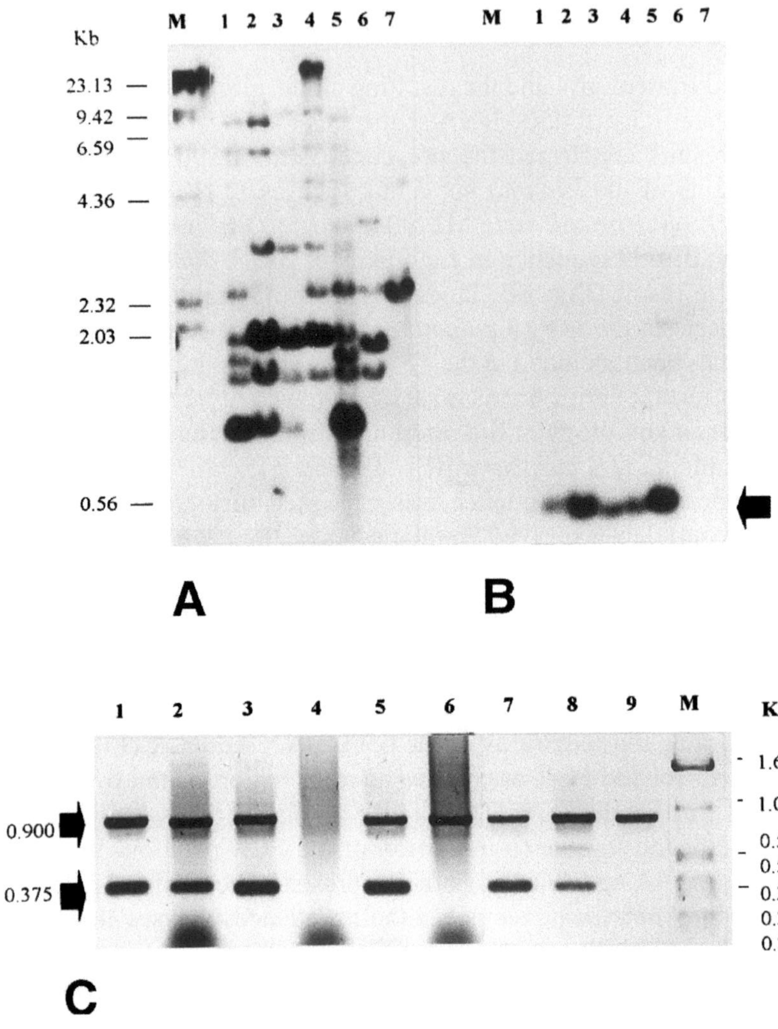

Fig. 6. Molecular markers specific of species of the *Leishmania (Viannia)* subgenus. The β-tubulin coding sequence and the β500-DNA sequence in *Leishmania* species of the *Viannia* subgenus. Total genomic DNA from several strains of New World *Leishmania* representative of both subgenera, *Leishmania* and *Viannia*, was digested with *Pst*I, fractionated on a 1% agarose gel, and after bidirectional transfer the membranes were hybridized independently, to the *Leishmania* β-tubulin ^{32}P–pLgβtub1 **(A)** and ^{32}P–pLgβ500 **(B)** probes. The reference strains of the *Viannia* subgenus include: lane *1*, *braziliensis* M2903; lane *4*, *panamensis* LS94. Lanes *2*, *3*, and *5* represent *Leishmania* isolates without previous identification. All isolates are identified as *L. (V.) braziliensis* strains. The *Leishmania* subgenus is represented by: lane *6*, *mexicana* BEL 21. Lane *7*, *colombiensis* CL500. The arrow indicates the position of the pLgβ500-DNA sequence. **(C)** Amplification of the coding region of the β-tubulin gene (0.90 kb) and ß500-DNA sequence (0.375 kb) from total genomic DNA of *Leishmania*

The β500-sequence probe is a very powerful molecular marker for the characterisation of the *Viannia* subgenus. Its absence in species of the *Leishmania* subgenus suggests a sequence divergence in the intergenic region of the tandem of the β-tubulin genes between the two *Leishmania* subgenera. Similar sequences to the β500 DNA, which is specific to species of the *Leishmania (Viannia)* subgenus, may be present in species of the *Leishmania (Leishmania)* subgenus. We are at present analyzing 5' and 3' sequences around the β-tubulin gene of *mexicana* and *donovani* species.

2. Materials

2.1. Parasites

2.1.1. Reference Strains

Leishmania strains studied are listed in **Table 1**. Most parasites have been characterized by several criteria and some represents WHO reference strains, which are routinely maintained in the laboratory in Schneider's *Drosophila* medium (GIBCO).

2.1.2. Growth and Maintenance of Parasites

2.1.2.1. Culture Medium

The cultivation, biological cloning, cryopreservation, and practice for handling *Leishmania* have been reviewed by Evans ***(25)***.

1. Schneider's insect medium (SIGMA, cat. no. S-9895).
   Dissolve the powdered medium in the distilled water, adjust the pH to 7.2 following the manufacturer's instructions, dispense into convenient screw-capped bottles, and autoclave. Storage culture at 4°C.
   The Schneider's insect medium supplemented with 10% fetal calf serum (GIBCO, cat. no. 10437-028), chloramphenicol at 20 μg/mL, and penicillin–streptomycin (5000 IU/mL–5000 μg/mL; GIBCO) (200 U/100 μg/mL) are added to the solution. *Leishmania* strains are cultured in 0.5 mL of medium, and for large preparation 100–200 mL are used routinely.
2. Biphasic blood agar.
   Solid phase: Mix 10.0 g of bacto agar (Difco), 3.0 g of NaCl, and 5.0 g of D(+)-

---

Fig. 6 ***cont.***
isolates and human biopsy. Lane *1*, *braziliensis* M2903; lane *2*, *panamensis* LS94; lane*s 3* and *5*, *Leishmania* isolates without previous identification; lane *4*, negative control (no DNA); lane *6*, *colombiensis* CL500; lane *7*, *guyanensis* M4147; lane *8*, human sample; and lane *9*, *major* P. The molecular marker (M) is represented by the 1-kb DNA ladder.

glucose in 0.50 L of distilled water. The mixture is sterilized by autoclaving. The agar is allowed to cool to about 55°C, then defibrinated rabbit blood added to a final concentration of 20%. Mix and dispense 1.0 mL into sterile culture tubes, and place the tubes in a sloped position until the agar solidified. Check the sterility of the medium by incubation at 37°C. Finally, store the tubes at 4°C.
Liquid phase: Sterile solution of 0.9% NaCl.
Both chloramphenicol at 20 μg/mL and penicillin–streptomycin (5000 IU/mL–5000 μg/mL; GIBCO)(200 U/100 μg/mL) are added to the solution.
Routinely, 0.2 mL of liquid phase are dispensed into a solid phase to subculture all *Leishmania* strains. This medium is suitable for isolation and maintenance of recent *Leishmania* isolates.

2.2. Genomic DNA Purification

All solutions and glassware should be autoclaved prior to use. Usually, individual solution stocks should be made with autoclaved water.

1. Washing buffer: 0.5 % glycerol,100 m*M* NaCl, 10 m*M* Tris-HCl, pH 7.5.
2. Lysis buffer: 0.5% Triton X-100, 50% glycerol, 100 m*M* $MgCl_2$, 50 m*M* Tris-HCl, pH 7.5, 300 m*M* NaCl.
3. Sodium dodecyl sulfate (SDS) buffer: 10 m*M* Tris-HCl, pH 7.4, 100 m*M* NaCl, 0.5% SDS, 1 m*M* EDTA.
4. ANE buffer: 10 m*M* Sodium acetate; 100 m*M* NaCl, 1 m*M* EDTA, pH 6 adjusted with glacial acetic acid.
5. Phenol–chloroform mixture: 300 mL of distillate phenol are added to a mix of 250 mL of buffer ANE plus 300 mL of chloroform and 0.05% (final) of 8-hydroxiquinoline; the mixture is stirred on a magnetic stirrer overnight. Finally, the mixture is left a room temperature for 2–3 h, the aqueous phase is removed and the solution is stored at 4°C in an amber bottle.
6. 7 *M* Ammonium acetate.
7. Autoclaved distillate water and MilliQ-water.
8. Ethanol: Analytical grade.
9. Six 30-mL glass centrifuge tubes (Corex) and 10 sterile microcentrifuge tubes with screw caps.
10. Two glass rods.
    Caution: For pipetting genomic DNA solution the tips are cut off at the end to avoid DNA shearing.

2.3. Restriction Enzyme Digestion

We selected a group of endonucleases based on the recognition sequence and their cutting frequency. The most informative endonucleases to analyze the tubulin gene region are shown in **Table 2**. Digestion conditions for a particular enzymes were those recommended by the manufacturer.

2.4. Electrophoresis and Southern Blotting of DNA

1. Agarose: Molecular Biology Certified Agarose Ultra Pure DNA Grade (Bio-Rad; cat. no. 162-0134).
2. 10× TBE buffer: 89 m*M* Trizma base, pH 8.3, 89 m*M* boric acid, 2.0 m*M* EDTA.
3. Sample buffer: 20% Sucrose, 100 m*M* EDTA, 0.025% bromophenol blue.
4. Ethidium bromide stock: 10 mg/mL.
5. Denaturation solution: 1.5 *M* NaCl, 0.5 *M* NaOH.
6. Neutralization solution: 0.1 *M* ammonium acetate, 0.02 *M* NaOH, pH 8.0.
7. Nylon membrane: One to two pieces of Hybond N+ from Amersham (RPN203B).
8. 20× SSC: 175.3 g of NaCl, 88.2 g of sodium citrate. 9. Whatman 3MM paper.

2.5. Cloning of the β-Tubulin Gene Region

The construction and the features of the genomic β-tubulin *Leishmania (V.) guyanensis* M4147clone, pLgβ4, and pLgβtub1 are described elsewhere ***(10)***. The cloning methodology, clone selection, and minipreparation of plasmid recombinant DNA is performed according to standard procedures ***(26)***.

1. Receptor cells *E. coli* DH5α.
2. Vectors: pUC18-*Bam*HI and pUC18-*Pst*I dephosphorylated under reaction conditions recommended by the supplier (Boehringer Mannheim).
3. Inserts: The 0.45-kb *Bam*HI fragment (Lgβtub1) and the 0.50-kb PstI fragment (Lgβ500), isolated from the pLgβ4 clone.

2.6. Labeling Probes

1. The multiprime DNA labeling kit.
2. [α-^{32}P]dCTP (3000 Ci/m*M*) from Amersham.

2.7. Hybridization

1. 1× SSC: 0.15 *M* NaCl, 0.015 *M* sodium citrate, pH 7.4.
2. 50× Denhardt stock solution: 5 g of Ficoll (Sigma), 5 g of polyvinilpyrrolidone (Sigma), 5 g of bovine serum albumin (fraction V), sterile distilled water to 0.5 L. The solution was filtered and storage at –20°C.
3. Hybridization buffer: 2× SSC/ 2× Denhardt, 2% SDS, 100 mg/mL of denatured sonicated calf thymus DNA (100 mg/mL).
4. Wash solution: 2% SDS/0.1% SDS.

2.8. Polymerase Chain Reaction

1. Source of DNA: Crude DNA, material from biopsy.
2. PCR buffer: PCR supermix (GIBCO-BRL, cat. no. 10572-014).
3. Primers: **Table 3**.

3. Methods

3.1. *Leishmania* Cultures

Promastigotes of *Leishmania* strains are grown at 25°C in Schneider's insect medium supplemented with 10–15% fetal calf serum (GIBCO), 20 μg/mL of chloramphenicol, and 100 μg/mL of ampicillin. Then parasites are harvested by centrifugation in the late log phase after 3–4 d in culture (approx 1×10^8 parasites/mL). Parasite preservation is carried out by dilution 1:1 (v/v) of parasites (log phase) with the same medium adjusted with 20% glycerol. The mixture is sequentially placed at 4°C for 2 h, at –10°C/2h, at –20°C/overnight, at –70°C/24 h and finally in liquid nitrogen.

3.2. Genomic DNA Purification (See Notes 1 and 2)

Different protocols of DNA isolation have been published; however, in our hands the DNA purification protocol described by Eresh et al. *(27)* is suitable for all analyses described in this chapter. The protocol given here is very simple, and we are able to obtain a good yield of DNA, adequately pure for most applications, with a high quality (*see* **Fig. 1**). The different steps of the protocol are as follows:

Cells are harvested by centrifugation at 5000g for 10 min. The pellet is resuspended in 10.0 mL of washing buffer and the centrifugation is repeated. Once the supernatant is discarded, the cells are gently lysed in 2 mL of lysis buffer at –10°C, using a glass rod and maintained for 10 min at the same temperature. Gentle resuspension and sequential addition of lysis buffer, starting with a small volume (approx 0.1 mL), is important. The resuspension procedure is done using a glass rod until homogeneity is reach. The suspension is centrifuged at 16,500g for 10 min at 0°C. Once centrifugation is completed the supernatant is discarded and 5 mL of SDS buffer is added to the pellet and incubated overnight at 37°C.

At this step one volume of a phenol–chloroform mixture (v/v) is added and mixed with the sample very gently by rotation of the tube several times. A second extraction is recommended. Then, carefully remove the aqueous phase (top) that contains the genomic DNA of high molecular weight and transfer the sample to a clean tube and add 0.5 volumes of 7 *M* ammonium acetate and mix by inversion. Subsequently, three volumes of ethanol are added to the sample. The DNA is recovered by centrifugation at 16,000g for 20 min. The ethanol is discarded and the tube are placed in an inverted position at –20°C for 10 min. The DNA is resuspended in 0.20–0.50 mL of MilliQ-water for further analysis.

3.3. Restriction Enzymes Digestion (See Note 3)

The digestion reaction of the *Leishmania* genomic DNA is prepared as follows:

Total genomic DNA (5 μg), prepared as described previously, is mixed with the appropriate restriction enzyme according to the manufacturer's conditions, in a final volume of 100 μL. The concentration of the enzyme will be approx 1 U/μg DNA. The sample(s) are incubated at 37°C overnight. To obtain complete digestion, the DNA sample is adjusted with fresh restriction buffer and enzyme, using no more than a quarter of the initial volume. The incubation is done at 37°C for 2 more hours. The sample is stored at 4°C until needed. Digestion of the PCR products is usually complete in 1–2 h. At the end of the digestion, the sample is size fractionated by gel electrophoresis. A typical reaction mixture contains:

10× buffer	10 μL (The appropriate one supply by the manufacturers)
DNA	5–10 μg
Enzyme	5 U
Pure water	appropriate to final volume of 100 μL

Digestion conditions of the PCR products were those recommended by the manufacturer. It was carried out over 1 h, with 1 U/μg of restriction enzyme at 37°C. For complete digestion, the DNA samples were adjusted with fresh restriction buffer and 1 U of enzyme was added and incubated again for another half-hour more. The sample was analyzed by electrophoresis on 2% agarose gel in TBE.

3.4. Electrophoresis and Southern Blotting of DNA

Different electrophoretic procedures and conditions are available, and it is difficult to offer general advice. Here we describe the general parameters involved in our protocol. It is as follows:

Agarose in appropriate concentration is melted in distilled water, the solution is placed at room temperature, and when it reached ~65°C, an appropriated volume of 10× TBE buffer was added at a final concentration of 1×. The gel is poured in a tray of a horizontal gel apparatus and 1× TBE is added until the gel surface is covered. The restriction fragments are separated by electrophoresis through 1% agarose gel at constant voltage (30 V) for 24 h. Previously, 0.1 volume of the sample buffer had been added to the DNA sample and loaded onto the gel. To analyze small fragments such as the products of the PCR–RFLP analysis, a 2% agarose gel is used and the electrophoresis is performed at 80 V for 4–5 h. The DNA fragments are visualized on a UV-transilluminator after staining the gel in an ethidium bromide solution (0.2 μg/mL) for 15 min. A photographic record is obtained in each case. The migration of the restriction fragments is compared to migration of markers run in the same gel, whose molecular weights are known.

There are different and excellent protocols to transfer DNA from agarose gel to filter. In our work, the DNA fragments are transferred bidirectionally onto a nylon membrane (e.g., Hybond N from Amersham) as described by Smith and Summers ***(28)***. The bidirectional method produces duplicate filters of the same gel which facilitates the comparison of two probes simultaneously. The different steps of the protocol include:

Soaking the gel in denaturation solution for 30 min to denature the DNA fragments. Then, rinse the gel with distilled water for 1 min and neutralize the gel in neutralization solution for 30 min.

At this time, two pieces of nylon membrane and four of Whatman 3MM paper, all the same size of the gel, are cut and immersed in the neutralizsation solution for 10 min. Previously, each membrane had been marked to identify the gel, its orientation, and the experiment.

The transfer sandwich system is assembled as follows:

Two sheets of Whatman paper are placed in a glass plate, over them one piece of nylon membrane (bottom membrane), previously marked on the edge with appropriate data. The gel is then laid underside uppermost on the transfer membrane, avoiding air bubbles. The second sheet of membrane is laid on top of the gel (top membrane) followed by two other pieces of Whatman paper. The sandwich is placed on the top of a paper towel (approx 10 pieces) laid on a glass plate, and then another packet of towel is put on top of the sandwich. Finally, a second plate of glass and two lead plates are placed on the top. Transfer is allowed to proceed overnight at room temperature. Disassemble the sandwich, soak the membranes in 2× SSC for 2 min, and bake for 30 min at 80°C between sheets of 3MM paper. The membranes are stored in a 3MM Whatman paper envelop at room temperature for further use.

Caution: All membranes must be handled and stored carefully at all times.

3.5. Cloning of the β-Tubulin Gene Region

The pLgβ4 plasmid is digested with the appropriate restriction enzymes, for example, *Bam*HI (pLgβtub1) or *Pst*I (pLgβ500), and after electrophoresis in a 1% agarose gel in TBE, the fragment of interest is retrieved from the agarose gel, purified by PrepA-gene purification system (Bio-Rad) and cloned in the pUC18 vector. Previously, the vector is digested with the same enzymes (*Bam*HI or *Pst*I) and dephosphorylated under reaction conditions recommended by the supplier (Boehringer Mannheim). The cloning methodology and clone selection are performed according to standard procedures using *E. coli* DH5α as receptor cells ***(10,26)***. Having identified the recombinant clones, pLgβtub1 and pLgβ500, we have characterized them in terms of the insert size, restriction map, and sequencing. Usually the amount obtained in a minipreparation of plasmid recombinant DNA is suitable for all analysis.

3.6. Labeling of DNA Probe

All probes were labeled with [α-^{32}P]dCTP by the random priming method using the multiprime DNA labeling kit from Amersham (RPN1604). The protocol followed is the one recommended by the manufacturer.

Mix the following reagents in a microcentrifuge tube at room temperature:

Template DNA (approx 0.1 μg)	
Reaction buffer	2.5 μL
Unlabeled dNTPs	2.0 μL (omitting those to be use as label)
[α-^{32}P] dCTP (3000 Ci/m*M*)	2.0 μL
Primers	2.5 μL
Water	10.0 μL (total volumen)
Enzyme	2.0 μL (2 U)
Final reaction volume	25.0 μL

The reaction mix is incubated at 37°C for 1 h.

3.7. Hybridization

Prehybridization and hybridization were carried out at medium stringency conditions (medium Cot), in hybridization buffer. The filter is placed in a plastic bag, 20 mL of hybridization buffer are added and incubated at 67°C for 1 h. Meanwhile, the radioactively labeled probe is heated in a boiling water bath for 5 min and chilled on ice. At the end of prehybridization time, the denature probe is added to the bag, which is then sealed and incubated at 67°C for 18 h. Then the filter is washed at medium stringency conditions in 2% SDS/0.1% SDS at 60°C with a minimum of four changes of buffer and exposed to X-ray film at –80°C.

3.8. Polymerase Chain Reaction (PCR)(See Note 4)

The starting material for the reaction can include purified crude DNA or cellular material processed with the QIAmp Tissue kit (QIAGEN, Chatsworth, CA). The PCR is performed in a final volume of 25 μL containing the cocktail of PCR-supermix, 10 pmol of primers (**Table 3**), and 5 ng of total genomic DNA. The reaction is carried out in a MJ Research PTC200 thermocycler, comprising 5 min of preincubation at 95°C followed by 35 cycles of 1 min at 95°C,

1 min at 55°C and 2 min at 72°C, with a final extension at 72°C for 5 min. The products are analyzed by electrophoresis on 1% or 2% agarose gel in TBE. Each assay contain a positive control, in which 100 fg of parasite DNA is included, and a negative control, in which no DNA is added. The reagents of a typical reaction to amplify the 0.9 kb fragment of the coding region of the β-tubulin genes contain:

PCR supermix	22.0 µL
Oligonucleotides (10 0 µ*M*)	1.0 µL (each)
DNA sample	1.0 µL

We have used the amplification of this fragment, as internal control of the PCR reaction in the β500 PCR assay at an annealing temperature of 55°C (*see* **Fig. 6C**). The PCR products are analyzed by electrophoresis in a 1–2% agarose gel prepared in TBE buffer.

4. Notes

Although the methodologies used in this study are relatively straightforward some assumptions and problems are important to comment. Different processes can explain the differences found among individuals in the pattern of DNA fragments changes in copy number in the tandem repeat sequence, base substitutions or insertion-deletion events, among others. Because the β-tubulin genes are a multigene family such as the ribosomal DNA (rDNA) we assume that in *Leishmania*, all genes in the β-tubulin tandem have the same sequence; in other words, they have evolved in a concerted manner. There is no evidence regarding this. Nevertheless, the restriction patterns found in the *Leishmania* reference strains are reproducible when recent field isolates of *Leishmania* are used. Moreover, there is the possibility that a second tandem of the gene may be present, at least in species of *Viannia* (*see* **Fig. 3A**).

A second assumption is to consider the base substitutions as the main process for the variation in the fragment restriction pattern of the tubulin gene. Base substitution can produce loss or gain of the sites for a particular enzyme. The complete β-tubulin gene sequence is available only for *L. (L.) amazonensis*, *L. (L.) major*, and *L. (V.) guyanensis* (Mendoza-León, *unpublished observations*), comparison of the restriction sites for different endonucleases support this assumption ***(21,22)***.

1. The DNA isolation and quality are very important in this kind of study; the methodology described in this work is very simple and a clean DNA of high molecular weight is obtained. The quality of the crude DNA is excellent, and it is not necessary to performed additional steps of purification. One important factor during

the DNA isolation is the sequential addition of lysis buffer in small volumes; resuspension of the parasites must be very gentle until homogeneity is reached. This procedure avoids the formation of clumps which decrease the yield of DNA due to the cell trapping caused by the genomic DNA released from the first lysed cells.

2. Another problem with the genomic DNA isolation protocol is the shearing of the DNA when a large number of extractions and long incubation time with chloroform-phenol mixture is used. To solve this problem we reduced the number of extractions to two and the incubation time with the mixture to 5 min.
3. RFLP analysis: The analytical and diagnostic purposes of this methodology are limited to the diagnostic restriction enzymes used because variation in the sequence(s) of interest might be undetected with other enzymes. It does not allow establishing differences among the strains or species of *Leishmania*. The differences found among *Leishmania* strains in the RFLP analysis of the β-tubulin gene may be analyzed from two perspectives. The first is by comparison of fragment mobility where we could not distinguish fragments produced during the endonuclease digestion, which could have comigrated. However, the restriction fragment patterns of the β-tubulin gene among *Leishmania* species showed sufficient differences to distinguish between New and Old World *Leishmania*. The second is by the comparison of the restriction sites; however, the complete β-tubulin gene sequence is available only for *L. (L.) amazonensis* ***(21)***, *L. (L.) major* (***22***) and *L. (V.) guyanensis* (A. Mendoza-León, *unpublished observations*). The comparison of these sequences was used to generate oligonucleotides and to select the restriction enzymes to improve a PCR–RFLP analysis. A direct analysis of nucleotide variation in PCR products may be improved by mutation detection techniques such as denaturing gradient gel electrophoresis (DGGE) or single-strand conformation polymorphism (SSCP).
4. PCR: We have optimized the β-500 PCR assay at different conditions of annealing temperature. We have improved the β-500 PCR assay at 55°C by dilution of the DNA sample; nevertheless, sometime the results are not the best. Routinely for clinical samples we use a temperature of 60°C for this assay. However, when biopsy material is analyzed we recommend running the reaction at a temperature of 65°C.

Acknowledgments

We are grateful to Drs. F.J. Tapia, M. Winkler, and A. Ramírez for critical review of the manuscript, and Jorge Rivas (IVIC) for photographic assistance. This work received financial support from the Venezuelan grants CDCH 03.10.4126.98, CDCH 03.33.3961.97, CONICIT S1-96001411 to A. M.-L. C. M. was financial supported for the Postdoctoral Program of CONICIT.

References

1. WHO (1990) Technical Report Series 793. *Control of the Leishmaniases*. Report of a WHO Expert Committee, Geneva. World Health Organisation.

2. Alvar E., J. P. (1997) Las Leishmaniasis: de la biología al control. Ed. Junta de Castilla y León.
3. Mendoza-León, A., Shaw, J. J., and Tapia, F. J. (1996) A guide for the cutaneous leishmaniasis connoisseur. In *Molecular and Immune Mechanisms in the Pathogenesis of Cutaneous Leishmaniasis* (Tapia, F. J., Cáceres-Dittmar, G., and Sánchez, M.A., eds.), Intelligence Unit Series. R. G. Landes Bioscience Publishers, TX, pp. 1–23.
4. Lainson, R. and Shaw, J. J. (1987) Evolution, classification and geographical distribution, in *The Leishmaniasis in Biology and Medicine*, vol 1 (Peters, W. and Killick-Kendrick, R., eds.), Academic Press, London, pp. 1–120.
5. WHO (1982) Recommendations. In *Biochemical Characterisation of Leishmania* (Chance, M. L. and Walton, B. C., eds.), UNDP/World Bank/WHO, pp. 263–276.
6. WHO (1984) Technical Report Series 701. *The Leishmaniases*. Report of a WHO Expert Committee, Geneva. World Health Organisation.
7. Eresh, S., de Bruijn, M. H. L., Mendoza-León, J. A., and Barker, D. C. (1995) *Leishmania (Viannia) lainsoni* occupies a unique niche within the subgenus *Viannia. Trans. R. Soc. Trop. Med. Hyg.* **89,** 231–236.
8. Mendoza-León, A., Havercroft, J. C., and Barker, D. C. (1995) The RFLP analysis of the β-tubulin gene region in New World *Leishmania. Parasitology* **111,** 1–9.
9. Mendoza-León, A. and Barker, D. C. (1996) Variation of a multigene family in New World *Leishmania*: the β-tubulin gene region. In *Molecular and Immune Mechanisms in the Pathogenesis of Cutaneous Leishmaniasis* (Tapia, F. J., Cáceres-Dittmar, G., and Sánchez, M. A., eds.), Intelligence Unit Series. R. G. Landes Bioscience Publishers, TX, pp. 107–127.
10. Luis, L., Ramirez, A., Aguilar, C. M., Eresh, S., Barker, D. C., and Mendoza-León, A. (1998) The genomic fingerprinting of the coding region of the β-tubulin gene in *Leishmania* identification. *Acta Trop.* **69**, 193–204.
11. Cleveland, D. W and Sullivan, K. F. (1985) Molecular biology and genetics of tubulin. *Ann, Rev. Biochem.* **54,** 331–365.
12. Fosket, D. E. and Morejohn, L. C. (1992) Structural and functional organisation of tubulin. *Ann. Rev. Plant Physiol. Plant Mol. Biol.* **43,** 201–240.
13. Landfear, S. M., McMahon-Pratt, D., and Wirth, D. (1983) Tandem arrangement of tubulin genes in the protozoan parasite *Leishmania enriettii. Mol. Cell. Biol.* **3,** 1070–1076.
14. Huang, P., Roberts, B. E., McMahon-Pratt, D. M., David, J. R., and Miller, J. (1984) Structure and arrangement of the beta tubulin genes of *Leishmania tropica. Mol. Cell. Biol.* **4,** 1372–1383.
15. Fong, D., Wallach, M., Keithly, J., Melera, P. W., and Chang, K.-P. (1984) Differential expression of mRNA for alpha and beta tubulin during differentiation of the parasite protozoan *Leishmania mexicana. Proc. Natl. Acad. Sci., USA* **81,** 5782–5786.
16. Spithill, T. W. and Samaras, N. (1985) The molecular karyotype *of Leishmania major* and mapping of alfa and beta tubulin gene families to multiple unlinked chromosomal loci. *Nucleic Acids Res.* **13,** 4155–4169.

17. Spithill, T. W. and Samaras, N. (1987) Genomic organisation, chromosomal location and transcription of dispersed and repeated tubulin genes in *Leishmania major*. *Mol. Biochem. Parasitol.* **24,** 23–37.
18. Nei, M. (1987). DNA polymorphism within and between populations. In *Molecular Evolutionary Genetics* (Nei, M. ed.), Columbia University Press, New York, pp. 254–286.
19. Brown, A. J. (1989) Population genetic at the DNA level: a review of the contribution of restriction enzymes studies. In *Oxford Surveys in Evolutionary Biology*, vol 6. Oxford University Press, pp. 207–239.
20. Dowling, T. E., Moritz, C., Palmer, J. D., and Rieseberg L. H. (1996) Nucleic acids III: Analysis of fragments and restriction sites. In *Molecular Systematics*, 2nd Edit. (Hillis, D. M., Moritz C., and Mable B. K., eds.), Sinauer Associates, Sunderland, MA, pp. 249–320.
21. Fong, D. and Lee, B. (1988) Beta tubulin gene of the parasitic protozoan *L. mexicana. Mol. Biochem. Parasitol.* **31,** 97–106.
22. Coulson, R. M., Connor, V., Chen, J. C., and Ajioka, J. W. (1996) Differential expression of Leishmania major beta-tubulin genes during the acquisition of promastigote infectivity. *Mol. Biochem. Parasitol.* **82,** 227–236.
23. Cupolillo, E., Medina-Acosta, E., Noyes, H., Momen H., and Grimaldi, G. Jr. (2000) A revised classification for *Leishmania* and *Endotrypanum*. *Parasitol. Today* **16**, 142–144.
24. Mendoza-León, A., Luis, L., Fernández, O., Cupolillo, E., and García, L. (2000). Molecular markers for species identification of the *Leishmania (Viannia)* subgenus, *Trans. R. Soc. Trop. Med. Hyg.* in press.
25. Evans, D. A. (1993) *In vitro* cultivation and biological cloning of *Leishmania*. In *Methods in Molecular Biology*, vol. 21. *Protocols in Molecular Parasitology* (Hyde, J. E., ed.) Humana Press, Totowa, NJ, pp. 29–41.
26. Sambrook, K. J., Fritsch, E. F., and Maniatis, T. (1989) *Molecular Cloning: A Laboratory Manual*, 2nd edit., Cold Spring Harbor Laboratory Press, Cold Spring Harbor, NY.
27. Eresh, S., Mendoza-León, A., and Barker, D. C. (1993) A small chromosome of *Leishmania (Viannia) braziliensis* contains multicopy sequences which are complex specific. *Acta Tropica* **55,** 33–46.
28. Smith, G. E. and Summers, M. D. (1980) The bidirectional transfer of DNA and RNA to nitrocellulose or diazobenzyloxymethyl-paper. *Analyt. Biochem.* **109,** 123–129.

9

Detection of Malignant Plasma Cells in the Bone Marrow and Peripheral Blood of Patients with Multiple Myeloma

Ross D. Brown and P. Joy Ho

1. Introduction

Multiple myeloma is a hematological malignancy characterized by an increased number of bone marrow plasma cells and the presence of high levels of a serum monoclonal immunoglobulin that is coded by a unique genetic sequence in the variable region of the immunoglobulin heavy chain gene. Morphological examination of bone marrow biopsy samples prior to therapy and serum immunofixation studies of the monoclonal immunoglobulin are usually sufficient to clearly diagnose this disease. However, after therapy when the number of malignant cells is reduced, it is not possible to differentiate between the small number of malignant and normal (polyclonal) plasma cells using traditional microscopy. In addition, while the malignant cells are generally found predominantly in the bone marrow, in many patients with progressive disease the malignant cells may spill over into the peripheral blood.

Flow cytometry has provided us with a tool to demonstrate the presence of plasma cells with light chain restriction, which infers clonality. Studies of peripheral blood stem cell harvests have shown that the contaminating plasma cells are predominantly polyclonal and thus not malignant *(1)*. However, the true identity of the malignant cells can be demonstrated only when gene probes are used, that correspond to the unique genetic sequence which is generated by recombination of the variable region and somatic hypermutation. This tumor-specific genetic signature does not change throughout the course of the disease *(2)*.

From: *Methods in Molecular Biology, vol. 179: Gene Probes: Principles and Protocols*
Edited by: M. Aquino de Muro and R. Rapley © Humana Press Inc., Totowa, NJ

Many laboratories have now performed polymerase chain reaction (PCR) with allele-specific oligonucleotides to identify the presence of the malignant clone in blood and bone marrow samples from patients with myeloma. One clinical application of this technique has been to demonstrate the presence of minimal residual disease after intensive therapy ***(3,4)***. However, the PCR method detects the presence of cell populations without distinguishing malignancy at the level of a single cell.

The technique of mRNA *in situ* hybridization (mRNA ISH) ***(5–8)*** was developed using patient—and tumor-specific probes to identify individual cells belonging to the malignant clone and to characterize other features of the malignant plasma cells at the single-cell level ***(9,10)***. Thus it has been shown that:

1. The malignant cells are present in the blood of all patients at concentrations of about 0.1–25% of the mononuclear cell fraction.
2. There is a direct correlation between disease activity and the number of malignant cells in blood.
3. Only a small proportion of the B cells ($CD19^+$) belong to the malignant clone.
4. Malignant $CD34^+$ cells either do not exist or are below the level of sensitivity of the assay **(9,10)**.

The mRNA *in situ* hybridization method we describe has been used with a number of different oligonucleotide and cDNA probes. The mRNA ISH technique is highly sensitive due to the high level of heavy and light chain mRNA present in the cytoplasm of plasma cells. Thus mRNA ISH using heavy or light chain probes has a good chance of overcoming any sensitivity problems, and strong staining of patient—and tumor-specific IgH chain mRNA can be achieved (*see* **Fig. 1**). The sensitivity of detection can be further increased by ISH-PCR, but this should not be necessary. The mRNA *in situ* hybridization technique consists of two major steps. First the CDR3 region of the immunoglobulin heavy chain gene of the malignant cell population is sequenced, from which the nongermline sequences are determined and biotinylated antisense oligonucleotide probes to these unique sequences are prepared ***(2,4,8–10)***. The protocol we outline describes the second stage which involves the *in situ* hybridization staining procedure for cytospin preparations of blood and bone marrow cells. The steps in this staining procedure are fixation and permeabilization, hybridization, detection and visualization.

2. Materials

1. TBS buffer 1: 0.1 *M* Tris-HCl, 1.0 *M* NaCl, 5 m*M* $MgCl_2$, pH 7.6. For 1 L: 12.1 g of Tris, 59.4 g of NaCl, 1 g of $MgCl_2$. Autoclave and store at room temperature.
2. TBS buffer 2: 0.1 *M* Tris-HCl, 0.1 *M* NaCl, 10 m*M* $MgCl_2$, pH 9.2. For 1 L: 12.1 g of Tris, 5.84 g of NaCl, 2.0 g of $MgCl_2$. Autoclave and store at room temperature.

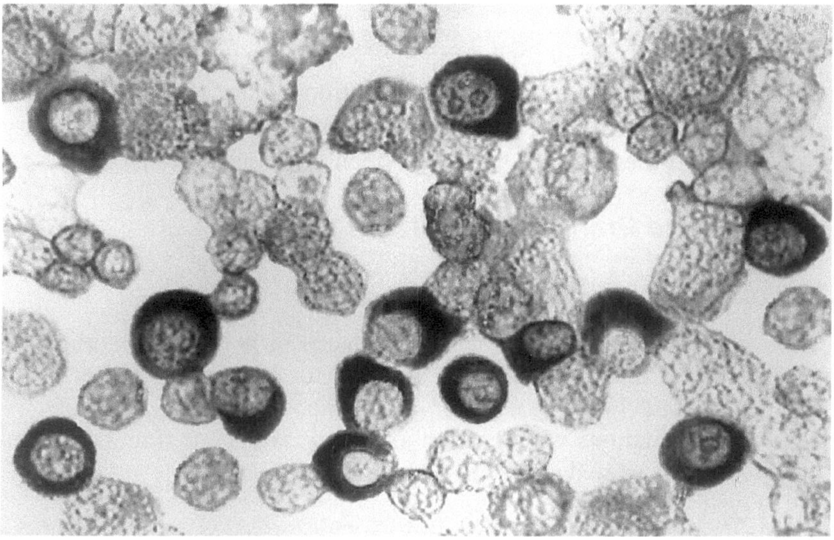

Fig. 1. *In situ* hybridization of bone marrow cells from a patient with multiple myeloma. Immunoglobulin heavy chain mRNA in the cytoplasm of plasma cells stained with probes that are patient and tumor specific.

3. Developing reagent A: 1 mL of TBS buffer 2, 10 µL of nitro blue tetrazolium (NBT), 10 µL of 5-bromo-4-chloro-3-indolyl phosphate (BCIP). Add NBT (Sigma) to buffer 2, mix well, and then add BCIP (Sigma) and mix. Prepare immediately before use. Store NBT and BCIP in the dark at –20°C.
4. Developing reagent B: 1 mL of TBS buffer 2, 10 µL of naphthol AS-MX phosphate (Sigma), 10 µL of Fast Red TR salt (Sigma), 10 µL of 0.5 *M* levamisol (Sigma). To prepare: Dissolve 20 mg of naphthol AS-MX phosphate in 1 mL of dimethylformamide (DMF) in a glass tube and dissolve 20 mg of Fast Red TR salt in 1 mL of distilled water. Add 10 µL of each solution to 1 mL of buffer 2 and mix well. Add 10 µL of 0.5 *M* levamisole and mix. Prepare immediately before use. Store naphthol AS-MX, Fast Red solutions, and levamisole at –20°C in the dark.
5. Hybridization buffer (for 10 mL): 1 g of dextran sulfate (mol wt 500,000; Sigma); 5 of mL formamide (Aldrich); 2.5 mL of 20× saline sodium citrate (SSC); 1 mL 10 m*M* EDTA; 0.5 mL of water; 1 mL of herring sperm DNA (2.5 mg/mL in 10 m*M* EDTA) (Roche). Preparation: Dissolve dextran sulfate in 20× SSC, add herring sperm DNA and water, then finally add formamide and mix well. Store at –20°C until required.
6. Phosphate buffered saline (PBS): For 1 L of 10× stock solution: 2.0 g of KCl, 80.0 g of NaCl, 11.5 g of Na_2HPO_4, 2.0 g of KH_2PO_4.
7. Sodium chloride-sodium citrate solution (SSC): For 1 L of 10× stock: 87.7 g of NaCl; 44.1 g of trisodium citrate.

3. Methods

This protocol does not describe the methods concerning extraction of DNA from the malignant plasma cells, amplification of the variable region of the *IgH* genes, DNA sequencing, probe design, and construction of biotinylated probes. Details of these techniques have been previously published ***(2,4,9,10)***.

3.1. Fixation and Prehybridization

3.1.1. Preparation of Slides

1. Slides are washed in ethanol and then placed in poly-L-lysine hydrobromide (0.005%) for 5 min. The slides are air-dried for 10 min and then at 37°C overnight. These slides may be used for up to 2 mo.
2. Cytospin preparations of cells from a primary sample or from cell culture are prepared on the treated slides.

3.1.2. Fixation with Ethanol and Acetic Acid

Several different fixation methods may be used. It is recommended that fixation with ethanol and acetic acid is attempted first. However, owing to the variability of the probes, one of the other methods (*see* **Notes 1** and **2**) should be tried if results are not as good as expected.

1. Fix air-dried slides of cells in 3:1 ethanol and acetic acid mixture for 30 min.
2. Place slides in 100% ethanol for 5 min.
3. Immerse slides in acetone for 5 min to extract lipids and to perforate the cell membrane.
4. Air-dry slides in an incubator at 37°C and store in a dust-free area.

3.2. Hybridization with Oligonucleotide Probes

1. Add 20–50 µL of biotin-labeled oligonucleotide probe (20–100 ng) to slides (*see* **Notes 4–7**).
2. Cover with coverslip.
3. Incubate at 42°C for 12 h or overnight but place in a moist chamber.

3.3. Posthybridization and Detection

1. Remove coverslips and rinse slides in two changes of 2× SCC–0.1% sodium dodecyl sulfate (SDS) at 42°C.
2. Slides are incubated for 15 min in 2× SSC–0.1% SDS at 42°C.
3. Incubate for 15 min in 0.4ß SSC–0.1%SDS at 42°C.
4. Slides are incubated for a further 15 min in 2× SSC–0.1% SDS at 42°C.
5. Incubate for 15 min in 0.4× SSC–0.1%SDS at 42°C.
6. Rinse slides in two changes of 0.4× SCC at room temperature.
7. Rinse slides in two changes of 2× PBS.

8. Slides are placed in two changes of TBS buffer for 1–5 min each.
9. The slides are drained and streptavidin (1:100 dilution in buffer 1 of DAKO ISH Detection kit – K0600) is applied to the slides for 10–20 min with occasional rocking.
10. Slides are rinsed in TBS buffer 1, and again drained.
11. Biotinylated alkaline phosphatase (1:100 dilution in buffer 1 of DAKO ISH Detection kit) is applied to the slides for 10 min with occasional rocking.
12. The slides are rinsed in TBS buffer 1.
13. The slides are drained and streptavidin (1:500 dilution in buffer 1 of DAKO ISH Detection kit–K0600) is applied to the slides for 10–20 min with occasional rocking.
14. Slides are rinsed in TBS buffer 1, and again drained.
15. Biotinylated alkaline phosphatase (1:500 dilution in buffer 1 of DAKO ISH Detection kit) is applied to the slides for 10 min with occasional rocking.
16. The slides are rinsed in TBS buffer 1.
17. The slides are incubated with two changes of TBS buffer 2 for 5 min.

3.4. Visualization

1. The slides are drained rapidly and excess developing reagent A or B (i.e., more than 50 μL) is added. At the same time a coverslip is placed over the slide to avoid bubbles. The excess developing reagent is blotted off and the slides are stored in a cool dark area.
2. The slides are viewed quickly under the microscope at 30-min intervals until the intensity of staining is satisfactory. The mRNA staining should be most intense in the cytoplasm.
3. The slides are washed several times with distilled water and counterstained (*see* **Note 8**) with Nuclear fast red (Sigma) for 5 min, washed with water, then metanil yellow for 3–4 min and washed with water.
4. Slides are mounted while wet using an aqueous mounting medium such as glycerol-gelatin.

4. Notes

1. All glassware should be baked at 180°C for 2 h, and all solutions should be made in autoclaved water and treated with 0.2% diethyl pyrocarbonate (DEPC). Salts should be RNase free. Wear gloves and use sterile disposable plasticware. See a technical manual, for example, Sambrook et al. ***(11)*** for additional details of basic molecular techniques.
2. There are several fixation methods that can be tried as alternatives if poor results are obtained. (a) Fixation with paraformaldehyde: Fix cells on slide in 4% paraformaldehyde dissolved in PBS for 30 min. Rinse slides with two changes of PBS for 5 min each. Wash slides with two changes of 0.25% Triton-X100 (Sigma) and 0.25% Nonidet P-40 (Sigma) in PBS for 5 min each to extract lipids and perforate the cell membrane. Rinse slides with two changes of PBS for 5 min each. Dip slides in 20% acetic acid (in water) for 5 min. Rinse slides in two

changes of distilled water for 5 min each. Place in 100% ethanol for 5 min. Air-dry at 37^{0}C in a dust-free area. (b) Fixation with HCTF (Histochoice Fixative – Astral): Fix slides in 1x HTCF (20× HTCF–alcohol–water = 1:4:15) for 30–40 min. Then place slides in 100% ethanol for 5 min, then in acetone for 5 min to extract lipids and perforate the cell membrane. Air-dry slides and store in a dust-free area. Cells must not be allowed to dry. This is likely to cause elevated levels of nonspecific binding, indicated by an overall blue background. BCIP/NBT positivity has a purple hue.

3. The oligonucleotides that we have successfully used were 18–24 mers. The optimum conditions may vary for different probes due to size and GC content. It is advisable to end-label with biotin during the construction of the probe. Longer probes will need higher temperatures for hybridization. Whenever there are several nongermline sequences, a cocktail of two or more different probes can be used to enhance the detection.
4. cDNA probes do not have as good penetration as oligonucleotide probes and require a longer incubation. One hundred microliters of photobiotin labeled probe (5 μg) and 20 μL of 10× SSC are placed in a sterile Eppendorf tube and heated at 90°C for 15 min to dissociate the double-stranded DNA into single strands and immediately chilled on ice to prevent reannealing. Cold hybridization buffer (900 μL) is then added and mixed well. The preparation is stored at –20°C until required. To the prepared slides is added the appropriate DNA probe in hybridization buffer (20–40 μL) and then completely covered with a coverslip. Slides are placed in a closed chamber containing absorbent paper moistened with water and incubated for 22–28 h at 40°C.
5. Gene sequences should be checked for uniqueness using an appropriate sequence database, for example, GenBank.
6. Control slides of cells from different patients hybridized with the same oligonucleotide probe should also be used. A probe of irrelevant specificity may be used as a negative control. Nonplasma cells act as internal negative control. Sense probes may be weakly positive.
7. Some cells have high levels of endogenous biotin to which streptavidin will bind nonspecifically. This can be overcome by using a different label. Digoxigenin labeled probes have been effective ***(12,13)***.
8. Counterstaining with nuclear stains may make interpretation difficult. This is especially true for hematoxylin. A light counter stain with Nuclear Fast Red is recommended. Alternatively Methyl green (Sigma) for 1 min can be used, washed with water or Mayer's hematoxylin for about 10 min, rinsed with water, and placed in a weak solution of ammonia solution for 1 min, and then rinsed in two changes of water.

Acknowledgments

The authors acknowledge the technical assistance of Xiao-Feng Luo and financial support from the Anthony Rothe Memorial Trust, Foundation IV and the Leo and Jenny Foundation.

References

1. Pope, B., Brown, R., Gibson, J., and Joshua, D. (1997) Plasma cells in peripheral blood stem cell harvests from patients with multiple myeloma are predominantly polyclonal. *Bone Marrow Transplant* **20,** 205–210.
2. Ralph, Q., Brisco, M. J., Joshua, D. E., Brown, R., Gibson, J., and Morley A. (1993) Advancement of multiple myeloma from diagnosis through plateau phase to progression does not involve a new B-cell clone: evidence from the immunoglobulin heavy chain gene. *Blood* **82**, 202–206.
3. Swedin, A., Lenhoff, S., Olofsson, T., Thuresson, B., and Westin, J. (1998) Clinical utility of immunoglobulin heavy chain gene rearrangement identification for tumour cell detection in multiple myeloma. *Br. J. Haematol.* **103,** 1145–1151.
4. Brisco, M. J., Tan, L. W., Osborn, A. M., and Morley, A. A. (1990) Development of a highly sensitive assay based on the polymerase chain reaction for rare B-lymphocyte clones in a polyclonal population. *Br. J. Haematol.* **75,** 163–167.
5. Lawrence, J. B. and Singer, R. H. (1985) Quantitative analysis of *in situ* hybridization methods for the detection of actin gene expression. *Nucleic Acids Res.* **13,** 1777–1799.
6. Gee, C. E. and Roberts, J. L. (1983) *In situ* hybridization histochemistry: a technique for the study of gene expression in single cells. *DNA* **2,** 157.
7. Pringle, J. H., Primrose, L., Kind C. N., Talbot, I. C., and Lauder, I. (1989) *In situ* hybridization demonstration of poly-adenylated RNA sequences in formalin-fixed paraffin sections using a biotinylated oligonucleotide poly d(T) probe. *J. Pathol.* **158,** 279–286.
8. Markovic, B., Kwan Y-L., Nicholls, E. M., Walsh, C., and Crouch, R. L. (1992) A sensitive method for the detection of poly-A tails of mRNA using a biotin labelled heteropolymer of dT:rA. *J. Pathol.* **167,** 369–373.
9. Brown, R. D., Luo, X-F, Gibson J., Morley A., Sykes, P., Brisco, M., and Joshua, D. E. (1995) Idiotypic oligonucleotide probes to detect myeloma cells by mRNA *in situ* hybridization. *Br. J. Haematol.* **90**, 113–118.
10. Joshua, D. E., Brown, R. D., Luo, X-F., and Gibson, J. (1996) Circulating clonal lymphocytes in myeloma determined by mRNA *in situ* hybridization. *Blood* **88,** 1125.
11. Sambrook, J., Fritsch, E. F., and Maniatis, T. (1989) *Molecular Cloning: A Laboratory Manual,* 2nd edit., Cold Spring Harbor Laboratory, Cold Spring Harbor, NY.
12. Kessler, C. (1990) The digoxigenin system: principle and applications of the novel nonradioactive DNA labelling and detection system. *BioTechnology Int.* **1990,** 183–194.
13. Holtke, H. J. and Kessler, C. (1990) Non-radioactive labeling of RNA transcripts in vitro with the hapten digoxigenin (DIG); hybridization and ELISA-based detection. *Nucleic Acids Res.* **18,** 5843–5851.

10

Detection of *Chlamydia trachomatis* by DNA Amplification

Peter Timms and Sarah Mathews

1. Introduction

The *Chlamydiae* are small, nonmotile Gram-negative bacteria that were originally thought to be viruses because of their size and their obligate dependence on host cells for growth. They are characterized by a unique biphasic developmental cycle involving an interconversion between the extracellular, infectious but metabolically inactive elementary body form (EB) and the intracellular, metabolically active but noninfectious reticulate body form (RB). Within an infected cell, the reticulate bodies multiply 200–500-fold by binary fission, resulting in a characteristic chlamydial inclusion, that often distends the cell. Of the four currently recognized chlamydial species (although Everett et al. ***(1)*** recently proposed a split of the *Chlamydiae* into two genera and nine species) only *Chlamydia trachomatis* and *C. pneumoniae* are common pathogens of humans. *C. trachomatis* is the more important pathogen of the two species for humans and it infects the mucosal surfaces of the cervix, urethra, rectum, nasopharynx, and conjunctiva. Cervical infections can ascend into the endometrium and the fallopian tubes, resulting in pelvic inflammatory disease, infertility, and ectopic pregnancy. Infection during pregnancy can adversely affect the newborn, leading to neonatal conjunctivitis and infant pneumonia. In males, *C. trachomatis* is a major cause of nongonococcal urethritis and ascending infections can lead to epididymitis. *C. trachomatis* infections are recognized worldwide as the most common bacterial sexually transmitted disease, and it is estimated that more than 50 million new cases of *C. trachomatis* infection occur annually ***(2)***. *C. trachomatis* infections are also known to increase the risk for human immunodeficiency virus (HIV) infection ***(3)***, and thus proper treatment of chlamydial infections can help delay the spread of HIV in some

From: *Methods in Molecular Biology, vol. 179: Gene Probes: Principles and Protocols*
Edited by: M. Aquino de Muro and R. Rapley © Humana Press Inc., Totowa, NJ

groups. The LGV serovars of *C. trachomatis* cause lymphogranuloma venerum, a sexually transmitted infection that involves the regional lymph nodes producing systemic involvement. In addition to genital infections, *C. trachomatis* is also the etiological agent of trachoma, the leading cause of preventable blindness in developing countries. The rapid and sensitive detection of infections due to *C. trachomatis* is therefore essential for the proper treatment of infected individuals and for the prevention of transmission of disease.

Perhaps the greatest challenge to the control of chlamydial disease is the fact that as many as 70–80% of women and up to 50% of men who are infected with *C. trachomatis* do not experience any symptoms ***(4,5)***. This results in a large reservoir of unrecognized, infected individuals who are capable of unknowingly transmitting the infection to their sexual partners. If these individuals could be easily detected (with reliable diagnostic tests), then they could be easily treated using the reliable and effective antibiotic treatments that are available for *Chlamydia*.

Diagnostic test methodologies for *C. trachomatis* have evolved significantly over the past 20 yr, from direct microscopy, through serology, cell culture, various antigen detection approaches, DNA probe hybridization, and most recently to nucleic acid amplification assay tests (NAATs). Although advantages exist for all the methods, it is now widely acknowledged that the NAAT format is the most sensitive and specific test for this important pathogen ***(6)***. A major advantage of the NAATs is their extra sensitivity, which means they can be used with noninvasive specimens such as first-catch urine samples, rather than the traditional vaginal swabs. The use of traditional cell culture approaches has all but disappeared and it is now realized that these tests are at best, 80% sensitive and perhaps as low as 50% ***(6)***.

Three types of NAATs have emerged: polymerase chain reaction (PCR, Roche Molecular Systems), ligase chain reaction (LCR, Abbot Laboratories), and transcription-mediated amplification (TMA, Gen-Probe Inc.). Both PCR and LCR generally target sequences in the cryptic chlamydial plasmid, primarily because it is present in 7–10 copies per elementary body, thus providing extra sensitivity for the assay. Overall, however, PCR has emerged as the most commonly used diagnostic approach. This is partly due to its versatility, making it not only well suited to commercial kits (e.g., Roche Amplicor) but also readily amenable to in-house test development.

As mentioned, the NAATs are well suited to use with a range of clinical specimens, including endocervical swabs and increasingly importantly, urines, allowing easier screening of males in particular. One challenge for the NAATs has been to avoid enzyme inhibition without adding unduly complicated sample processing steps. While preparation methods do vary considerably, most involve a centrifugation step to concentrate the particulate material, followed by some type of cell lysis procedure, with or without protein digestion.

In many specimens low numbers of chlamydial organisms are present, and this, combined with the differing levels of sensitivity provided by the various diagnostic assays, means that it is not uncommon for different results to be obtained with different assays, even on the same specimens. Because of these discrepancies, it is often considered desirable to include some type of quality assurance step in *C. trachomatis* testing. Therefore, it is usually recommended that all positive results are confirmed, particularly for low prevalence populations, asymptomatic patients, and those patients for whom a false-positive result would have adverse effects. As has been mentioned, NAATs are now considered to be significantly more sensitive than other types of test procedures. For this reason, the only suitable confirmatory assay for confirming a NAAT is another, independent NAAT. Because the gene target used for most *C. trachomatis* PCR assays is now the cryptic chlamydial plasmid, it is common to target the chlamydial major outer membrane protein (*MOMP*) gene (*ompA*) for confirmatory testing. This still has some shortcomings, however, as the plasmid is present in multiple (10) copies per cell whereas the *MOMP* gene is single copy. Nevertheless, PCR confirmation targeting the *ompA* gene is probably the most reliable means of confirming a plasmid polymerase chain reaction (PCR)-positive specimen.

Although the method used to detect the *Chlamydia* is important, collecting an adequate specimen for testing is even more critical. Numerous reports have confirmed that the sensitivity and specificity of the test depend directly on the quality of the specimen that is collected ***(7,8)***. Because chlamydiae are obligate intracellular pathogens, the objective of any specimen collection procedure should be to include the hosts cells that harbor the organism. This usually means taking an abrasive sample. The use of NAATs has partially reduced the reliance of optimal specimens, primarily because the assays are so sensitive that they can detect lower numbers of EBs in the sample. For this reason, urine samples have become more common samples for use with NAAT test formats. Nevertheless, the specimen must be adequate or the resultant sensitivity of the assay will be compromised.

The basic steps in a procedure to detect *C. trachomatis* by PCR are as follows: (1) specimen collection, (2) sample preparation, (3) DNA amplification, and (4) hybridization capture and EIA detection of product.

2. Materials

1. Target gene—PCR primers: The DNA target for amplification is a 207-base pair (bp) segment of the genetically conserved *C. trachomatis* cryptic plasmid: for-

ward primer = 5'-GGGATTCCTGTAACAACAAGTCAGG-3'; reverse primer = 5'-CCTCTTCCCCAGAACAATAAGAACAC-3' (*see* **Notes 1** and **2**). PCR primers can be commercially synthesized and minimum purification either by desalting or reverse-phase column treatment is desirable. PCR primer stocks should be aliquotted on receipt and stored frozen at –80°C to avoid excessive freeze–thaw cycles, which can cause oligonucleotide degradation. Always check a new batch before using all of the current batch.

2. Proteinase K solution: Prepare a stock solution of proteinase K at 20 mg/mL in TE buffer and store frozen in aliquots at –70°C. Use at a final concentration of 50 µg/mL.
3. PCR buffer: 10 m*M* Tris-HCl, pH 8.3, 50 m*M* KCl, 2.5 m*M* $MgCl_2$, 0.01% gelatin. Stable at –20°C for at least 6 mo.
4. dNTPs: Available as a stock PCR mix from Roche. They can be stored for up to 12 mo at –20°C and should be diluted in water to 2 m*M* for use
5. PCR master mix: 200 µ*M* (each) of dATP, dCTP, dGTP, and DIG-dUTP; 2 U *Taq* polymerase (AmpliTaq); 2.5 m*M* $MgCl_2$, 1 µ*M* forward and reverse target primers in PCR buffer.
6. PCR tubes: Use 0.2—or 0.6-mL PCR tubes. Some tubes have thinner plastic walls than others and therefore can be cycled between the various PCR temperatures more rapidly, making the total PCR time shorter.
7. Capture probe: The capture probe has a sequence of 5'-CATAGCACTATAGAACTCTGCAAGCC-3' and is 3'-end labeled with biotin (by standard labeling procedures). Resuspend the lyophilized capture probe in TE buffer to a stock concentration of 100 µ*M*. This stock can be stored at –20°C for up to 6 mo. For use, dilute the stock in water to 7.5 p*M* and use 2 µL per 50 µL of PCR hybridization reaction.
8. 20× SSC: 20× Saline sodium citrate (SSC) is made up of 3*M* NaCl and 0.3 *M* Na citrate at pH 7.0.
9. Hybridization solution : 1× SSC solution.
10. Streptavidin-coated microtiter plates: Suitable plates are available from NEN Life Sciences (cat. no. NEF711). Plates are stable at 4°C for up to 3 mo.
11. Hybridization wash solution: 1× SSC solution.
12. Anti-DIG horseradish peroxidase conjugate: Use a 1:2500 dilution of commercial anti-DIG HRP conjugate (Boehringer Mannheim) in 100 m*M* Tris-HCl, pH 7.5, 150 m*M* NaCl, 0.1% Tween-20, and 10 mg/mL bovine serum albumin (BSA). Stable at 4°C for up to 3 wk.
13. EIA wash solution: Phosphate-buffered saline containing 1% Tween-20.
14. TMB substrate: TMB Chromagen (TMB-9060, PanBio Pty, Brisbane, Australia).
15. Stop solution: 1 *M* H_2SO_4.

3. Methods

3.1. Sample Preparation

3.1.1. Endocervical and Conjunctival Swabs

1. Place the swab into 1 mL of PCR buffer supplemented with 200 μg of proteinase K per milliliter and 1% Tween-20.
2. Incubate at 55°C for 1 h.
3. Use 5–20 μL of this treated specimen directly in the PCR reaction.

3.1.2. Urine Specimens

1. Use first-catch urine samples only, males or females (*see* **Note 3**). Check the urine for the presence of a precipitate. If present, warm at 37°C for 30 min to dissolve any crystals. Vortex-mix thoroughly for 10 s.
2. Centrifuge 1 mL of urine specimen at 12,000 rpm in a microfuge for 20 min at room temperature. Discard the supernatant.
3. Resuspend the pellet in 100 μL of PCR buffer supplemented with 200 μg of proteinase K per milliter and 1% Tween-20.
4. Incubate at 55°C for 1 h.
5. Use 5–20 μL of this treated urine specimen directly in the PCR reaction.

3.2. DNA Amplification

1. Add 5 μL of treated specimen to 45 μL of PCR master mix in a 0.2 mL of thin-walled PCR tube.
2. Perform the PCR amplification as follows:

Cycle 1:	95°C for 5 min plus 60°C for 1 min
Cycles 2–30:	95°C for 30 s plus 60°C for 60 s (29 cycles)
Hold:	Hold at 72°C for the next step

3.3. Hybridization Capture and EIA Detection of Product

1. Add 10 μL of PCR product to 2 μL of biotin-labeled capture probe in 38 μL of 1× SSC hybridization solution (total volume of 50 μL).
2. Heat to 95°C for 5 min, then quench on wet ice for 2 min (*see* **Note 4**).
3. Transfer the whole volume to individual wells of the streptavidin-coated microtiter plate and incubate at 37°C for 20–30 min.
4. Discard the supernatant and rinse the well three times with room temperature enzyme-linked immunosorbent assay (ELISA) wash solution (200 μL per well per wash cycle).
5. Add 200 μL of anti-DIG horseradish peroxidase conjugate and incubate at 37°C for 30 min.
6. Wash as described in **step 4**.
7. Add 150 μL of TMB substrate, and allow the color to develop for 10–60 min.

8. Add an equal volume of stop solution and read the optical density (OD) at 450 nm.
9. An OD of >0.20 is considered positive (*see* **Notes 5** and **6**).

4. Notes

1. PCR target: The target chosen for amplification is the cryptic chlamydial plasmid. A 207-bp fragment located 195 bp downstream from the unique *Bam*HI restriction site is targeted ***(9)*** in this PCR assay. This target has the advantage that it is present in multiple copies ***(7–10)*** per cell, hence providing extra sensitivity ***(10,11)***. The plasmid also provides a degree of specificity because it is found in all *C. trachomatis* strains but not in the other human chlamydial species, such as *C. pneumoniae*.
2. Prevention of PCR contamination: Several procedures can be used to prevent or at least significantly reduce, PCR contamination, including: (a) decontaminate work surfaces and equipment with 1 *N* HCl or 10% hypochlorite solution; (b) wear gloves; (c) use dedicated pipets, either positive displacement pipets or pipets with filter tips (aerosol barrier); (d) physically separate the areas used for specimen preparation, PCR setup (inside a Biohazard cabinet), and ELISA detection; (e) use the Roche AmpErase system. AmpErase contains the enzyme uracil *N*-glycosylase (UNG) which recognizes and catalyzes the destruction of deoxyuridine-containing DNA, but not thymidine-containing DNA. Deoxyuridine is not present in microbial DNA, but is present in all Amplicor amplicons owing to the use of deoxyuridine triphosphate (in place of thymidine triphosphate) as one of the dNTPs in the PCR mastermix. Thus only amplicons will contain deoxyuridine, making them susceptible to destruction by AmpErase prior to amplification of the target DNA. In addition, negative controls should be included with each run. Ideally, one negative control should be included for each 5–10 test specimens (preferably interspersed between the test samples, rather than done as a single group at the end of the run).
3. Elimination of specimen inhibitors: Urine specimens should be collected into a polypropylene container that does not contain preservatives. Specimens are stable at room temperature for 24 h but should be stored at 4°C if held longer and tested within 4 d of collection. Uric acid is a known inhibitor of PCR and should be removed or at least diluted out by the centrifugation procedure. Do not resuspend the urine pellet in smaller volumes, as this may cause inhibition problems.
4. Hybridization capture: Heat denaturation of the PCR product is convenient and easy. However, it is also possible to use alkali denaturation and acid–HEPES buffer neutralization. Mix equal volumes of PCR reaction and 0.8 M NaOH solution and leave at room temperature for 5 min. Add an equal volume of pH 7.0 1 *M* HEPES buffer and use immediately for the probe capture.
5. Assay sensitivity: In theory, PCR can detect as few as one single target molecule. In practice however, a sensitivity of around 10–100 chlamydial particles is more likely, but is still several logs more sensitive than non-NAATs. Detection of the PCR product by hybridization capture and ELISA is also usually found to be 10–

100 times more sensitive than ethidium bromide-stained agarose gel detection. Sample adequacy can be confirmed by coamplification of β-globin or β-actin gene sequences along with the target *C. trachomatis* DNA.

6. Confirmatory testing: Because of differences between tests, there will often be some discrepancy between results. Because of the extra sensitivity provided by the NAATs, it is essential to use a second NAAT as a confirmatory test. A second PCR targeting a separate gene (such as *ompA*) is probably the best approach. The method described above can be modified slightly to target a segment of the *ompA* gene using the CT0005-CT06 primers described by Bobo et al. *(12)* (CT0005 = 5'-GATAGCCAGCACAAAGAGAGCTAA-3'; CT06 = 5'-CTTTGTTTTCGACCGTGTTTTGCAAACAGATGTGAA-3').

References

1. Everett, K., Bush, R., and Andersen, A. (1999) Emended description of the order *Chlamydiales*, proposal of *Parachlamydiaceae* fam. nov. and *Simkaniaceae* fam. nov., each containing one monotypic genus, revised taxonomy of the family *Chlamydiaceae*, including a new genus and five new species, and standards for the identification of organisms. *Int. J. Syst. Bacteriol.* **49,** 415–440.
2. Krul, K. (1995) Closing in on *Chlamydia. CAP Today* **9,** 1–20.
3. Laga, M., Manoka, A., Kivuvu, M., Malele, B., Tuliza, M., and Gowman, J. (1993) Non-ulcerative sexually transmitted diseases as risk factors for HIV-1 transmission in women: results from a cohort study. *AIDS* **7,** 95–102.
4. Schachter, J., Stoner, E., and Moncada, J. (1983) Screening for chlamydial infections in women attending family planning clinics. *West. J. Med.* **138,** 375–379.
5. Stamm, W. and Cole, B. (1986) Asymptomatic *Chlamydia trachomatis* urethritis in men. *Sex. Transm. Dis.* **13,** 163–165.
6. Schachter, J. (1999) Which test is best for *Chlamydia*? Curr. *Opin. Infect. Dis.* **12,** 41–45.
7. Howard, C., Friedman, L., Leete, J., and Christensen, M. (1991) Correlation of the percent of positive *Chlamydia trachomatis* direct fluorescent antibody detection tests with the adequacy of specimen collection. *Diagn. Microbiol. Infect. Dis.* **14,** 233–237.
8. Schwebke, J., Stamm, W., and Handsfield, H. (1990) Use of sequential enzyme immunoassay and direct fluorescent antibody tests for the detection of *Chlamydia trachomatis* infections in women. *J. Clin. Microbiol.* **28,** 2473–2476.
9. Loeffelholz, M., Lewinski, C., Purohit, A., Herman, S., Buonagurio, D., and Dragon, E. (1992) Detection of *Chlamydia trachomatis* in endocervical specimens by polymerase chain reaction. *J. Clin. Microbiol.* **30,** 2847–2851.
10. Mahoney, J., Luinstra, K., Sellors, J., Jang, D., and Chernesky, M. (1992) Confirmatory polymerase chain reaction testing for *Chlamydia trachomatis* in first-void urine from asymptomatic and symptomatic men. *J. Clin. Microbiol.* **30,** 2241–2245.
11. Mahoney, J., Luinstra, K., Sellors, J., and Chernesky, M. (1993) Comparison of

plasmid—and chromosome-based polymerase chain reaction assays for detecting *Chlamydia trachomatis* nucleic acids. *J. Clin. Microbiol.* **31,** 1753–1758.

12. Bobo, L., Coultee, R., Yolken, T., Quinn, T., and Viscidi, R. (1992) Diagnosis of *Chlamydia trachomatis* cervical infection by detection of amplified DNA with an enzyme immunoassay. *J. Clin. Microbiol.* **28,** 1968–1973.

11

Isolation and Identification of *Campylobacter* Species by Use of Selective Enrichment, Nucleic Acid Amplification Methods, and Gene Probes

Alex van Belkum, Nicole van den Braak, Leen-Jan van Doorn, and Hubert Endtz

1. Introduction

Bacteria belonging to the genus *Campylobacter* are a well-recognized cause of food-related gastrointestinal infections. These infections are usually self-limiting and constitute substantial morbidity. Mortality, fortunately, is low. Deaths directly attributable to *Campylobacter* infections in the United States are estimated to be 120–360 cases per year. Detection and identification of these bacteria is of limited clinical importance, as the time needed for detection usually exceeds the duration of the symptoms of the disease. However, sometimes more significant sequelae of infection, such as the Guillain-Barré syndrome (GBS), can be documented. The GBS is characterized by severe and sometimes fatal polyneuropathy. Owing to molecular mimicry between *Campylobacter* surface antigens and some human gangliosides, the immunological response elicited by the bacteria may eventually result in serious nerve degeneration as the consequence of autoimmune reactivity. Therefore, the availability of highly specific and sensitive methods for detection of antecedent *Campylobacter* infection for patients with GBS is mandatory and of clinical importance. High-quality tests may provide an early marker for a more protracted and severe course of the disease. Rapid methods may help to identify and treat GBS patient with a poor prognosis. Consequently, several molecular-diagnostic assays have been developed over the years in addition to conventional microbiological procedures.

From: *Methods in Molecular Biology, vol. 179: Gene Probes: Principles and Protocols*
Edited by: M. Aquino de Muro and R. Rapley © Humana Press Inc., Totowa, NJ

Since the late 1980s, the identification of *Campylobacter* spp. has been performed by a large number of phenotypic methods ***(1–3)***. During subsequent years various elegant nucleic acid-mediated methods, either alone or in combination with short-term enrichment culture, were developed. Most of these methods relied on the use of specific DNA probes suited for species identification after isolation ***(4–15)***. Others combined direct amplification of *Campylobacter* spp. DNA from clinical or environmental material with probe-mediated confirmation of the nature of the amplicons ***(16–28)***. These individual studies provided the experimental basis of several of the technical procedures outlined in one of the forthcoming sections.

We describe here the procedure that we developed and use in our laboratory for the detection and identification of *Campylobacter* spp. in fecal samples of GBS patients and in other samples where low numbers of Campylobacters are expected. Owing to the extended time between the preceding gastrointestinal infection and the onset of GBS, the excretion of *Campylobacter* cells is thought to be extremely low at the moment of analysis. For this reason, highly sensitive enrichment procedures are required, because mere detection of microbial DNA is insufficient. Because the pathogenesis of GBS is incompletely understood, isolation of viable strains is extremely important for determining the etiological role of *Campylobacter* spp. in the development of this paralytic syndrome. Once isolated, bacterial strains are specified using DNA amplification and hybridization assays. For this purpose, we use a recently developed test system based on the nucleotide sequence of a *Campylobacter* gene, that encodes a putative GTPase ***(14,29)***. As with all GTPases, the protein contains several semiconserved GTP-binding sites designated G-1, G-3, and G-4 ***(30)***. The *Campylobacter* GTPase represents a special family of GTPases with two adjacent GTP-binding domains. Degenerate PCR primers, based on the G-1 and G-3 sites of the first GTP-binding domain of the *C. jejuni* GTPase gene, allow amplification of a 153-basepair long fragment from *C. jejuni, C. coli, C. lari,* and *C. upsaliensis* as well ***(31)***. Because the sequence of the fragments is species specific, polymerase chain reaction (PCR) amplification of the elements can be combined with reversed oligonucleotide hybridization to define the species nature of a putative *Campylobacter* isolate. The combination of recovery by selective enrichment and molecular identification has led to a laboratory protocol that is highly sensitive and specific and which was instrumental for establishing one of the world's largest collections of GBS-associated *C. jejuni* strains ***(32)***.

2. Materials

2.1. Isolation

1. Physiological salt solution: Dissolve 9 g of NaCl in 1 L of distilled water. Fill 50-mL bottles, autoclave, and store at room temperature (stable for at least a year after sterilization).
2. Cellulose acetate filter: Pore size 0.65 µm (Sartorius, Goetingen, Germany).
3. Campylobacter Thioglycolate Broth (CTB): The medium is prepared using compounds sold by Beckton Dickinson (Cockeysville, USA) and contains per liter: 17 g of pancreatic digest of casein, 3 g of papaic digest of soybean meal, 6 g of dextrose, 2.5 g of sodium chloride, 0.5 g of sodium thioglycolate, 1.6 g of agar, 0.25 g of L-cystein, 0.1 g of sodium sulfite, 2 mg of amphotericin-B, 15 mg of cephalotin, 5 mg of trimethoprim, 10 mg of vancomycin, and 2500 U of polymyxin-B. The medium is filter sterilized and stored in 10-mL bottles at 4°C. The medium can also be purchased from Becton Dickinson (Cockeysville, USA; cat. no. 4321748).
4. Rotterdam Curaçao Selective Broth (RCSB): 18.5 g of brain heart infusion medium (Difco 0037-01-6, Detroit, USA) is suspended in 450 mL of distilled water and sterilized at 121°C for 15 min. The solution is left to cool down to 50°C, where after 50 mL of sterile horse serum is added. Subsequently, cefoperazone, amphotericin B, teicoplanin (CAT) supplement (Oxoid SR174, Hampshire, England) and Preston selective supplement (Oxoid CM689, Hampshire, England) of which standard portions are dissolved in 2 mL of water each, are added. The medium is stored in 10-mL portions up to 2 mo at 4°C.
5. *Campylobacter* agar (CAT agar): This medium is prepared from blood-free *Campylobacter* agar base (Oxoid CM739, Hampshire, England) and *Campylobacter* selective supplement (CAT, Oxoid SR174, Hampshire, England; see also the preceding for amounts and procedures).
6. Butzler agar: This is a blood agar medium prepared from Columbia Agar Base (Oxoid CM331, Hampshire, England) and 5–7% defibrinated horse blood (Becton Dickinson, Cockeysville, USA). Added is a standard portion of Butzler Selective Supplement (Oxoid SR85, Hampshire, England).
7. Mueller Hinton-based blood agar medium: Mueller Hinton medium (38 g of Mueller Hinton base [Difco, Breda, The Netherlands] per liter of distilled water with pH 7.3) either with or without 5% defibrinated sheep blood. Both the Mueller Hinton and the MH blood agar plates can be purchased ready-to-use (Biotrading, Mijdrecht, The Netherlands).
8. Glycerol broth: This medium contains per liter: 1 g of KH_2PO_4, 4 g of K_2HPO_4, 5 g of NaCl, 15 g of proteose peptone no. 3 (Difco) and 150 mL of filter-sterilized glycerol. The medium is sterilized at 121°C for 15 min and dispensed in portions of 1.2 mL.

2.2. Microbiological Identification

1. Oxidase test: Difco (Breda, The Netherlands).
2. Catalase test: 10% Hydrogen peroxide solution.
3. Camp-ID indoxyl acetate test: Mast Diagnostics, Mercyville, UK (contains hazardous agents).
4. Sodium hippurate solution 1%: Dissolve 0.15 g of sodium hippurate (Sigma) in 15 mL of distilled water and divide in 0.5-mL portion in Eppendorf tubes. Store at –20°C.
5. Ninhydrin: Dissolve 3.5 g of ninhydrin (2,2-dihydroxy-1,3-indianedione; Sigma, Zwijndrecht, The Netherlands) in 100 mL of acetone–butanol (1:1); store in the dark at room temperature (hazardous agent).
6. Nalidixic acid disc NA30: Oxoid, Hampshire, England.
7. Cephalotin disc KF30: Oxoid, Hampshire, England.

2.3. Molecular Identification

2.3.1. DNA Isolation

1. Phosphate buffered saline (PBS): Dissolve one tablet (Oxoid BR14a, Hampshire, England) in 100 mL of distilled water and autoclave at 115°C for 15 min. The pH of the solution is 7.3 and it can be stored for 1 yr at 4°C.
2. Isolation buffer: 10 m*M* Tris-HCl, pH 8.0, 5 m*M* EDTA, 0.1% SDS, 0.1 mg/mL of proteinase K (Boehringer Mannheim, Mannheim, Germany).

2.3.2. Polymerase Chain Reaction

1. PCR buffer: 10 m*M* Tris-HCl, pH 8.3; 50 m*M* KCl, 4.0 m*M* $MgCl_2$, 200 m*M* dNTPs. This buffer is manufactured by Perkin Elmer as a 10× concentrated stock solution and should be stored at –20°C. It is possible but not mandatory to use separate buffer and dNTP stock solutions.
2. *Taq* polymerase: AmpliTaq Gold (Perkin Elmer, Gouda, The Netherlands).
3. Primers: Forward primers GTP-F1 (5'-biotin-GGiAARCCAAATGTiGGiAARTC-3') and GTP-F1D (5'-biotin-GAAAACCAAATGTYGGCAAATC-3') are based on the G-1 GTP binding site (consensus $G(X)_4GKS/T$; X means any amino acid). The sequence of a single reverse primer GTP-R1 (5'-biotin-CTYTCAT CRAGiCCiCCRCTATC-3') is based on the G-3 GTP-binding site (consensus DXXG).

2.3.3. Line Probe Assay

1. Probe strips: Probes Cj 41 (CTTTTTAATAGAATGGCAAGACAAAG), Cj 413 (CTTTTTAATAGAATGGCAAGACAA) and Cj 443 (TTCATATTCATTCAA AAAAAGC-CATGCTT) are specific for *C. jejuni*. Probes Cc 453 (ATCAAG TTTATTTAACAGA-ATGGCAAG) and Cc 473 (CCAATAAAACAGAATGG TTTATAAATTC) are specific for *C. coli*. Probes Cl 492-2 (GACTTGCAA GAAARCGYAT), Cl 493 (TAGACTTGCAAAA-AAACGCAT), Cl 50-2

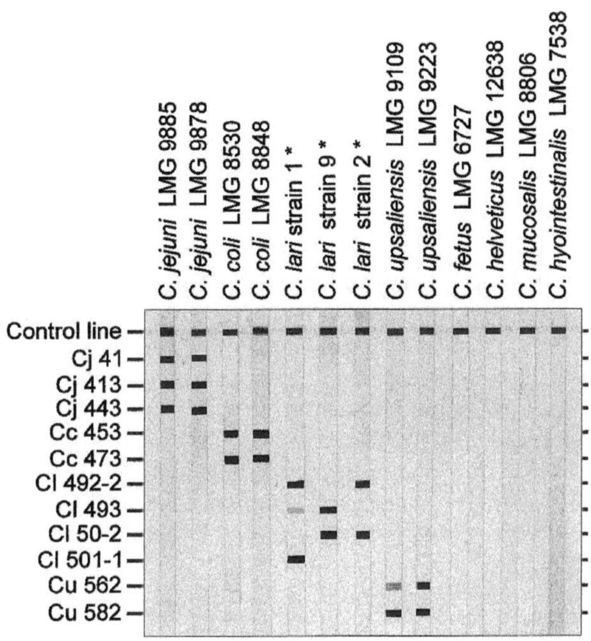

Fig. 1. LiPA for various *Campylobacter* isolates. Nine different *Campylobacter* strains isolated from stool samples were analyzed. The LiPA strips show the specificity of hybridization of the labeled PCR products toward the immobilized oligonucleotides. Note that especially the *C. lari* amplicons show differential specificity toward the probes. This is caused by allelic variability in the target PCR region. Cj, *C. jejuni*; Cc, *C. coli*; Cl, *C. lari*; Cu, *C. upsaliensis*.

(ACATAAGYGGRACYACAAG), and Cl 501-1 (ATATTAGC-GGCACAACAAG) hybridize to *C. lari* amplimers. Probes Cu 562 (TAATCGCATAGCA-AGGCAA) and Cu 582 (TATCAATGGTAAAGAAGCCTT) specifically recognize *C. upsaliensis* (Y = C or T; S = G or C; R = G or A; i = inosine). The oligonucleotide probes were enzymatically provided with a poly d(T) tail. Probes were immobilized as parallel lines on nitrocellulose membrane strips (*see* **Fig. 1**). The top line is a positive control and contains biotinylated *Campylobacter* DNA.

2. Denaturation solution: 400 m*M* NaOH (Sigma), 10 m*M* EDTA.
3. Hybridization buffer: 45 m*M* sodium citrate, 450 m*M* sodium chloride, 0.1% sodium dodecyl sulfate (SDS).
4. Rinse solution: 30 m*M* sodium citrate, 300 m*M* sodium chloride, 0.1% SDS.
5. Alkaline phosphatase–streptavidin conjugate: A dilution of 150 mU/mL is made in rinse solution. The enzyme complex is purchased from Boehringer Mannheim (Mannheim, Germany).
6. Substrate buffer: 100 m*M* Tris-HCl, pH 9.5, 100 m*M* NaCl.

7. Substrate: As a substrate for staining the complexes of probe, PCR product and alkaline phosphatase–streptavidin conjugate, 5-bromo-4-chloro-3-indolyl phosphate (BCIP) and nitroblue tetrazolium (NBT) are employed. A dilution of the stocks, bought from Boehringer Mannheim, is made by adding 35 μL of BCIP and 45 μL of NBT to 10 mL of substrate buffer.

3. Methods

3.1. Isolation

1. A fecal smear is directly inoculated on a CAT agar and Butzler agar (d 1) (*see* **Note 1**).
2. Approximately 0.5 g of feces is suspended in 10 mL of physiological salt solution. Six drops of this suspension are placed on the cellulose acetate filter that is placed on a Mueller Hinton blood agar plate.
3. The filter is incubated for 30–60 min at 37°C. Subsequently, the filter is removed.
4. All agar plates are incubated in a microaerobic atmosphere for 3 × 24 h at 37°C. The presence of 10% H_2 has been shown to be beneficial for some strains. However, one should take care not to exceed the low-explosive level of H_2.
5. Plates are examined visually every day.
6. On day 1, 1 g of feces is suspended in RCSB and CTB enrichment medium as well.
7. These media are incubated microaerobically at 37°C for 18 h with shaking.
8. A 100-μL portion is inoculated onto CAT and Butzler agar plates and a Mueller Hinton blood agar plate.
9. Plates are incubated for another 72 h and analyzed as described previously. Besides the fecal sample itself, portions of the CTB and RCSB enrichment suspensions are stored at –20°C for possible follow-up experiments.
10. *Campylobacter* spp. grow as gray, wet-shining colonies that may show signs of swarming. All colonies with different morphology are identified.
11. Five different colonies, preferably from different media, are stored at –80°C in glycerol broth (*see* **Note 2**).

3.2. Microbiological Identification

1. Suspect colonies, meeting the morphological criteria mentioned previously, are investigated further.
2. Gram staining is routinely implemented (*see* **Note 3**).
3. The oxidase test is performed according to the instructions of the manufacturer. When oxygen is formed upon suspension of bacteria in the H_2O_2 solution, the reaction is considered positive.
4. A suspect colony is immersed in the indoxyl acetate solution with a cotton swab and incubated at 37°C for 30 min. When the solution stains dark blue, the test is positive. *C. lari* NTCC 11352 is used as negative control. *C. jejuni* NCTC 11351 is used as positive control.
5. Susceptibility toward the antibiotics nalidixic acid and cephalotin is determined by disk diffusion tests on blood agar.

6. Suspect colonies are suspended in the hippurate solution at an optical density of 1 McFarland. The tube is incubated at 37°C for 2 h. Subsequently, the ninhydrin solution is added to completely cover the surface of the hippurate solution. If after 10 min a purple coloration is visible at the interface of the suspension and the ninhydrin solution, the reaction is positive. *C. jejuni* NCTC 11351 and *C. lari* NCTC 11352 are used as positive or negative control, respectively (*see* **Note 4**).

3.3. Molecular Identification

3.3.1. DNA Isolation

1. Bacterial DNA is isolated from liquid bacterial cultures or a suspension of bacteria scraped from solid growth media. Bacteria are harvested by centrifugation and suspended in 2 mL of sterile PBS.
2. Cells are pelleted by centrifugation for 2 min at 12,000 g, suspended again in 400 µL of 10 m*M* Tris-HCl, pH 8.0, 5 m*M* EDTA, 0.1% SDS, 0.1 mg/mL of proteinase K and incubated for at least 2 h at 55°C.
3. Proteinase K is inactivated for 10 min at 95°C and the lysate is clarified by centrifugation for 5 min at 12,000 g.
4. The supernatant is diluted 1:100 in sterile water and used directly for PCR (*see* **Note 5**).

3.3.2. Polymerase Chain Reaction

1. PCR reaction mixtures (50 µL) are prepared in 1x PCR buffer. One reaction mixture contains 1.5 U of AmpliTaq Gold DNA polymerase and 70 pmol of each primer.
2. PCR comprises 40 cycles of 30 s at 94°C, 45 s at 60°C, and 45 s at 74°C. Prior to cycling, the samples were heated at 94°C for 9 min to activate the AmpliTaq Gold DNA polymerase (*see* **Note 6**).
3. Approximately 10 ng of bacterial DNA, as estimated by gel electrophoresis, or 2 µL of the 1:100 diluted bacterial lysate is used as template for amplification.

3.3.3. Line Probe Assay (LiPA)

1. PCR product (10 µL), containing biotin moieties at the 5' end of the primers, is denatured by adding 10 µL of denaturation solution.
2. After 10 min, 2 mL of prewarmed (37°C) hybridization buffer is added to the denatured PCR products, a LiPA strip is submerged in the tray and incubation proceeds at 50°C ± 0.5°C for 1 h.
3. The strips are washed twice for 30 s and once for 30 min at 50°C with 2 mL of hybridization solution. Following this stringent wash, three additional short rinsing periods with the rinse solution are included.
4. The strips are incubated with 2 mL of alkaline phosphatase–streptavidin conjugate for 30 min at room temperature.
5. Strips are washed twice with 2 mL of rinse solution and once with 2 mL of substrate buffer.

6. Substrate (2 mL) is added and the strips are incubated for 30 min at room temperature. The reaction is stopped by aspiration of the substrate solution and addition of 2 mL of distilled water.
7. The reverse hybridization is usually performed in an AutoLiPA‰ machine (Innogenetics, Gent, Belgium) and results are interpreted visually. The procedure can also be performed manually in a shaking water bath (*see* **Note 7**).

4. Notes

1. In The Netherlands, bacteria belonging to the species *Campylobacter* are in the risk class 2, requiring C-I laboratory facilities. Gloves should be worn during the manipulation of fecal samples, and it is advised to do the same during the analysis of suspect colonies.
2. If feces samples cannot be processed instantaneously, the material can be stored at 4°C for 18 h prior to inoculation on the diverse media. A positive growth control sample should be included (e.g., *C. jejuni* NCTC 11351) during all incubation steps. The remainder of the faecal sample is suspended in glycerol broth and stored in portions at –20°C. This material may in a later stage be used for molecular testing.
3. *C. jejuni* present as a slightly curved Gram-negative rod upon microscopic examination.
4. *C. jejuni* should be positive in the katalase-, oxidase-, hippurate-, and indoxyl acetate test and should be susceptible for nalidixic acid and resistant for cephalotin. It is important to stress, however, that the emergence of resistance against the fluoroquinolones may lead to problems with respect to the phenotypic identification to the species level. Note that exceeding the incubation period in the hippurate assay may cause a false-positive result. A negative oxidase assay need not be conclusive in the decision whether or not *Campylobacter* is involved, as oxidase-negative *C. jejuni* strains have been described before. Several additional pitfalls can be documented, but since the emphasis of this chapter should be on the molecular diagnostics, the reader is referred to the standard microbiology literature for more detail.
5. It has to be mentioned that the simple and straightforward DNA isolation protocol described here can be replaced by alternative procedures, which generate more pure DNA samples. Disadvantage of more complex DNA isolation procedures is the enhanced risk for contamination that is detrimental to the subsequent PCR tests. The protocol given here is simple as well as rapid and as such to be preferred over more lengthy ones. Another advantage of the protocol used is that heat-inactivated digests can be stored or transported at ambient temperatures for at least 6 mo.
6. For PCR, both of the two forward primers need to be combined with the reverse primer: in both cases the primer pairs show limited degeneracy, but useful PCR products will be generated. Probe length and composition were adapted to allow specific hybridization of all probes at a single hybridization condition. Multiple

probes are included per species to enhance the quantitative and qualitative specificity of the test.

7. Inclusion of multiple probes was essential to completely cover the entire spectrum of isolates belonging to a given species: there is significant allelic variation in the target genes involved (*see* **Fig. 1**). No cross-reactivity of the probes exists with amplified DNA from the species *C. fetus, C. helveticus, C. mucosalis*, or *C. hyointestinalis*. It has to be emphasized that setting the correct hybridization temperature is an essential reaction condition. It has been demonstrated before that the assay is highly specific, approaching the 100% value ***(31)***. Finally, once a *Campylobacter* culture is obtained, molecular species identification can be performed in a single working day.

5. General Remarks

We are aware of the fact that our complete protocol of both classical and molecular detection and identification of *C. jejuni* is labor-intensive and complex. However, the procedure outlined in detail in the preceding guarantees a highly sensitive and species-specific approach for the isolation of Campylobacters from fecal samples derived from GBS patients and other specimens where low numbers of *Campylobacter* organisms are to be expected such as food and environmental samplings. In our opinion, only scrutinously and cautiously performed protocols for the isolation and identification of Campylobacters will facilitate the elucidation of the pathogenesis and epidemiology of diseases caused by *Campylobacter* species.

References

1. Nachamkin, I. (1997) Microbiologic approaches for studying *Campylobacter* species in patients with Guillain-Barré Syndrome. *J. Infect. Dis.* **176,** S106–114.
2. Endtz, H. P., Ruys, G. S. H. M., Zwinderman, A. H., Van der Reyden, T., Biever, M., and Mouton, R. P. (1991) Comparison of six media, including a semi-solid agar, for the isolation of various *Campylobacter* species from stool specimens. *J. Clin. Microbiol.* **29,** 1007–1010.
3. Nachamkin, I. and Barbagallo, S. (1990) Culture confirmation of *Campylobacter* spp. by latex agglutination. *J. Clin. Microbiol.* **28,** 817–818.
4. Picken, R. N., Wang, Z., and Yang, H. L. (1987) Molecular cloning of a species-specific DNA probe for *Campylobacter jejuni*. Mol. Cell. *Probes* **1,** 245–259.
5. Korolik, V., Coloe, P. J., and Krishnaipillai, V. (1988) A specific DNA probe for the identification of *Campylobacter jejuni*. *J. Gen. Microbiol.* **134,** 521–529.
6. Chevrier, D., Larzul, D., Megraud, F., and Guesdon, J. L. (1989) Identification and classification of *Campylobacter* strains by use of nonradioactive DNA probes. *J. Clin. Microbiol.* **27,** 321–326.
7. Zhou, Y. T, and Wang, S. Q. (1989) Application of a biotin-labelled DNA probe to detect *Campylobacter*. *Zentralbl. Bakteriol.* **272,** 186–190.

8. Thorne, G. M., Macone, A., and Goldmann, D. A. (1990) Enzymatically labelled nucleic acid (NA) probe assays for detection of *Campylobacter* spp in human faecal specimens and in culture. *Mol. Cell. Probes* **4,** 133–142.
9. Tenover, F. C., Carlson, L., Barbagallo, S., and Nachamkin, I. (1990) DNA probe culture confirmation assay for identification of thermophilic *Campylobacter* species. *J. Clin. Microbiol.* **28,** 1284–1287.
10. Olive, D. M., Johny, M., and Sethi, S. K. (1990) Use of an alkaline phosphatase-labeled synthetic oligonucleotide probe for detection of *Campylobacter jejuni* and *Campylobacter coli*. *J. Clin. Microbiol.* **28,** 1565–1569.
11. Taylor, D. E. and Hiratsuka, K. (1990) Use of non-radioactive DNA probes for detection of *Campylobacter jejuni* and *Campylobacter coli* in stool specimens. *Mol. Cell. Probes* **4,** 261–271.
12. Gebhart, C. J., Murtaugh, M. P., Lin, G. F., and Ward, G. E. (1990) Species-specific DNA probes for *Campylobacter* species isolated from pigs with proliferative enteritis. *Vet. Microbiol.* **24,** 367–379.
13. Cudjoe, K. S., Kapperud, G., and Olsvik, O. (1991) Identification of thermotolerant *Campylobacter* species from poultry using an enzyme-labelled oligonucleotide DNA probe. *Int. J. Food Microbiol.* **12,** 271–274.
14. Giesendorf, B. A., Van Belkum, A., Koeken, A., Stegeman, H., Henkens, M. H., Van der Plas, J., et al. (1993) Development of species-specific DNA probes for *Campylobacter jejuni, Campylobacter coli,* and *Campylobacter lari* by polymerase chain reaction fingerprinting. *J. Clin. Microbiol.* **31,** 1541–1546.
15. Quentin, R., Chevrier, D., Guesdon, J. L., Martin, C., Pierre, F., and Goudeau, A. (1993) Use of nonradioactive DNA probes to identify a *Campylobacter jejuni* strain causing abortion. *Eur. J. Clin. Microbiol. Infect. Dis.* **12,** 627–630.
16. Giesendorf, B. A. J., Quint, W. G. V., Henkens, M. H. C., Stegeman, H., Huf, H. A., and Niesters, H. G. M. (1992) Rapid and sensitive detection of *Campylobacter* spp. in chicken products by using the polymerase chain reaction. *Appl. Environ. Microbiol.* **58,** 3804–3808.
17. Uyttendaele, M., Schukkink, R., Van Gemen, B., and Debevere, J. (1995) Detection of *Campylobacter jejuni* added to foods by using a combined selective enrichment and nucleic acid sequence-based amplification (NASBA). *Appl. Environ. Microbiol.* **61,** 1341–1347.
18. Hernandez, J., Alonso, J. L., Fayos, A., Amoros, I., and Owen, R. J. (1995) Development of a PCR assay combined with a short enrichment culture for detection of *Campylobacter jejuni* in estuarine waters. FEMS Microbiol. Lett. **127,** 201–206.
19. Manzano, M., Pipan, C., Botta, G., and Comi, G. (1995) Polymerase chain reaction assay for detection of *Campylobacter coli* and *Campylobacter jejuni* in poultry meat. *Zentralbl. Hyg. Umweltmed.* **197,** 370–386.
20. Winters, D. K. and Slavik, M. F. (1995) Evaluation of a PCR based assay for specific detection of *Campylobacter jejuni* in chicken washes. *Mol. Cell. Probes* **9,** 307–310.
21. Waegel, A. and Nachamkin, I. (1996) Detection and molecular typing of *Campylobacter jejuni* in fecal samples by polymerase chain reaction. *Mol. Cell. Probes* **10,** 75–80.

22. Oyofo, B. A., Mohran, Z. S., El-Etr, S. H., Wasfy, M. O., and Peruski, L. F. Jr. (1996) Detection of enterotoxigenic *Escherichia coli*, *Shigella* and *Campylobacter* spp. by multiplex PCR assay. *J. Diarrhoeal Dis. Res.* **14,** 207–210.
23. Harmon, K. M., Ransom, G. M., and Wesley, I. V. (1997) Differentiation of *Campylobacter jejuni* and *Campylobacter coli* by polymerase chain reaction. *Mol. Cell. Probes* **11,** 195–200.
24. Winters, D. K., O'Leary, A. E., and Slavik, M. F. (1997) Rapid PCR with nested primers for direct detection of *Campylobacter jejuni* in chicken washes. *Mol. Cell. Probes* **11,** 267–271.
25. Linton, D., Lawson, A. J., Owen, R. J., and Stanley, J. (1997) PCR detection, identification to species level, and fingerprinting of *Campylobacter jejuni* and *Campylobacter coli*direct from diarrheic samples. J. Clin. Microbiol. **35,** 2568–2572.
26. Ng, L. K., Kingombe, C. I., Yan, W., Taylor, D. E., Hiratsuka, K., Malik, N., and Garcia, M. M. (1997) Specific detection and confirmation of *Campylobacter jejuni* by DNA hybridization and PCR. *Appl. Environ. Microbiol.* **63,** 4558–4563.
27. Metherell, L. A., Logan, J. M., and Stanley, J. (1999) PCR-enzyme-linked immunosorbent assay for the detection and identification of *Campylobacter* species: application to isolates and stool samples. *J. Clin. Microbiol.* **37,** 433–435.
28. Waage, A. S., Vardund, T., Lund, V., and Kapperud, G. (1999) Detection of small numbers of *Campylobacter jejuni* and *Campylobacter coli* cells in environmental water, sewage and food samples by a seminested PCR assay. *Appl. Environ. Microbiol.* **63,**1636–1643.
29. Van Doorn, L. J., Giesendorf, B. A., Bax, R., Van der Zeijst, B. A., Vandamme, P., and Quint, W. G. V. (1997) Molecular discrimination between *Campylobacter jejuni, Campylobacter coli, Campylobacter lari* and *Campylobacter upsaliensis* by polymerase chain reaction based on a novel putative GTPase gene. *Mol. Cell Probes* **11,** 177–185.
30. Bourne, H. R., Sanders, D. A., and McCormick, F. (1991) The GTPase superfamily: conserved structure and molecular mechanism. *Nature* **349,** 117–127.
31. Van Doorn, L. J., Verschuuren-van Haperen, A., Burnens, A., Huysmans, M. B., Vandamme, P., Giesendorf, B. A., et al. 1999. Rapid identification of thermotolerant *Campylobacter jejuni, Campylobacter coli, Campylobacter lari*, and *Campylobacter upsaliensis* from various geographic locations by a GTPase-based PCR-reverse hybridization assay. *J. Clin. Microbiol.* **37,** 1790–1796.
32. Endtz, H. Ph., Ang, C. W., Van den Braak, N., Duim, B., Rigter, A., Price, L. J., et al. (2000) Molecular characterization of *Campylobacter jejuni* from patients with Guillain-Barré and Miller Fisher syndrome. *J. Clin. Microbiol.* **38,** 2297–2301.

12

Diagnosis of Human Papillomavirus Using *In Situ* Hybridization and *In Situ* Polymerase Chain Reaction

Gerard J. Nuovo

1. Introduction

In situ hybridization is the only DNA—or RNA-based molecular biology based test that allows for the direct correlation of the results with the histologic and cytologic features of the sample. The DNA/RNA extraction that precedes filter hybridization (hybrid capture or Southern blot) hybridization and the polymerase chain reaction (PCR) precludes this type of analysis. In order for the *in situ* hybridization to detect a given DNA or RNA target, there must be at least 10–20 copies of the target per cell. In comparison, Southern blot hybridization and the hybrid capture assay can detect one DNA or RNA target per 100 cells. PCR is even more sensitive if the hot start maneuver is employed; under these conditions, it can detect one DNA or RNA target per 100,000 cells. It is evident that *in situ* hybridization is a relatively insensitive test. A reflection of this relative insensitivity is seen in occult or latent infection by a virus where the copy number is low. In such situations, the virus is rarely detected by *in situ* hybridization even though it was detected by either PCR or filter hybridization (***1–6***). As stressed in this chapter, this is actually an advantage of *in situ* hybridization over the more sensitive assays. *In situ* hybridization detects only productive infection by human papillomavirus (HPV). It will not detect the low copy subclinical or "incidental" infection of HPV in the setting of a normal Pap smear (**Fig. 1**). When one realizes that incidental HPV infection can occur in 15–20% of normal Papanicolaou (Pap) smears, as determined by the hybrid capture or PCR assays, it is evident that it is advantageous for *in situ* hybridization NOT to detect the virus, given the realization that most concur there is no reason to clinically treat HPV in the absence of any pathology. This

From: *Methods in Molecular Biology, vol. 179: Gene Probes: Principles and Protocols*
Edited by: M. Aquino de Muro and R. Rapley © Humana Press Inc., Totowa, NJ

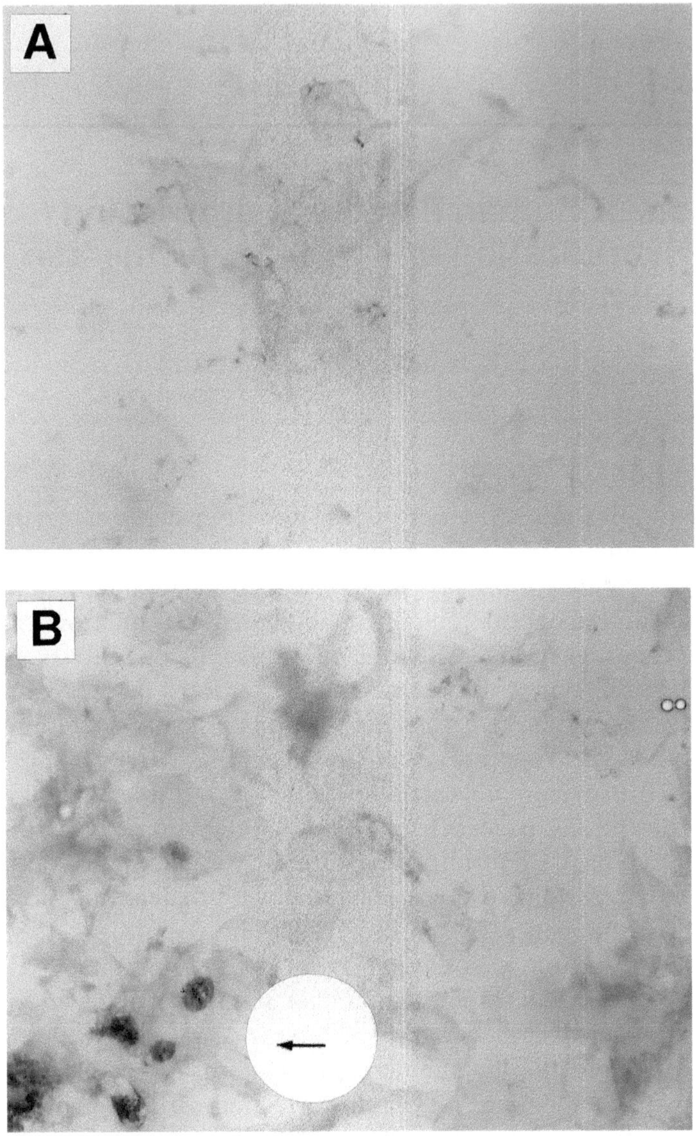

Fig. 1. Detection of HPV DNA in Pap smears. **(A)** A Pap smear that was within normal limits; HPV DNA was detected using the hybrid capture system (not shown). However, HPV DNA was not detected by *in situ* hybridization (**A**, pink is negative and blue is positive). **(B)** Specimen is from a woman with a Pap smear diagnosis of ASCUS. HPV DNA was detected by the hybrid capture system and *in situ* hybridization (**B**). The *in situ* test detects only HPV associated with productive infection and, thus, clinical disease.

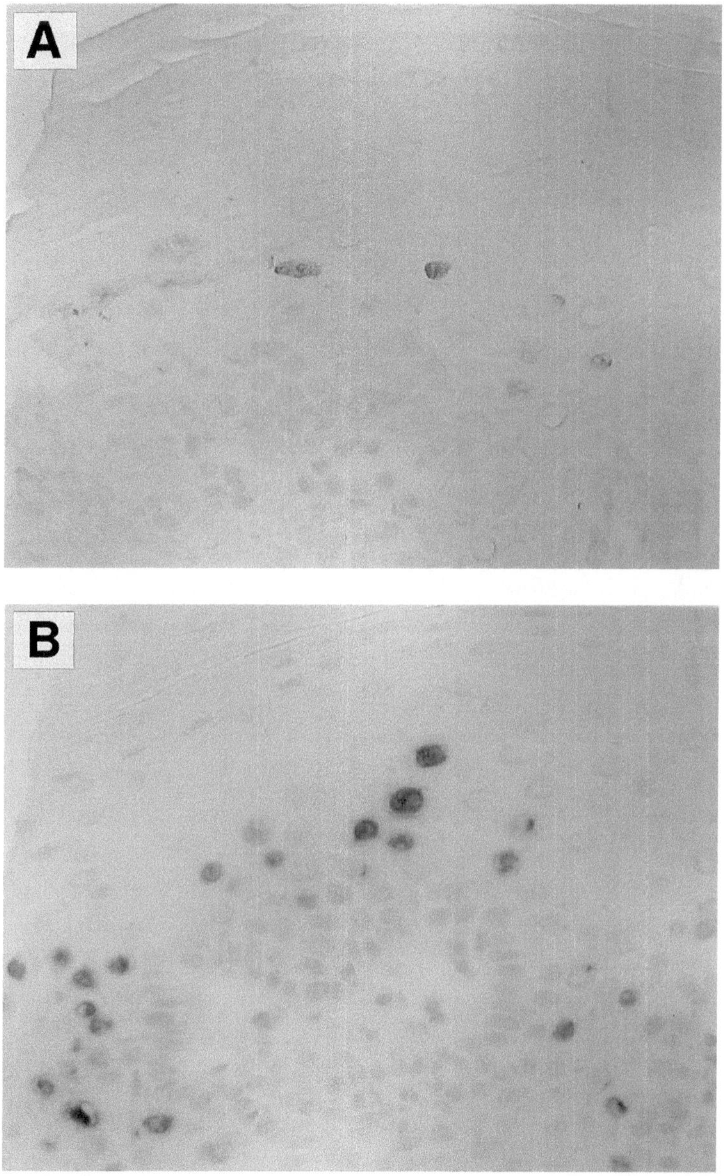

Fig. 2. Comparison of PCR *in situ* hybridization vs *in situ* hybridization with catalyzed signal amplification. This nongenital wart contained HPV 2. Note that the virus is present in scattered cells as detected by *in situ* hybridization with CSA **(A)**. More cells were detected in the serial section from the same group of cells if PCR *in situ* hybridization was employed **(B)**.

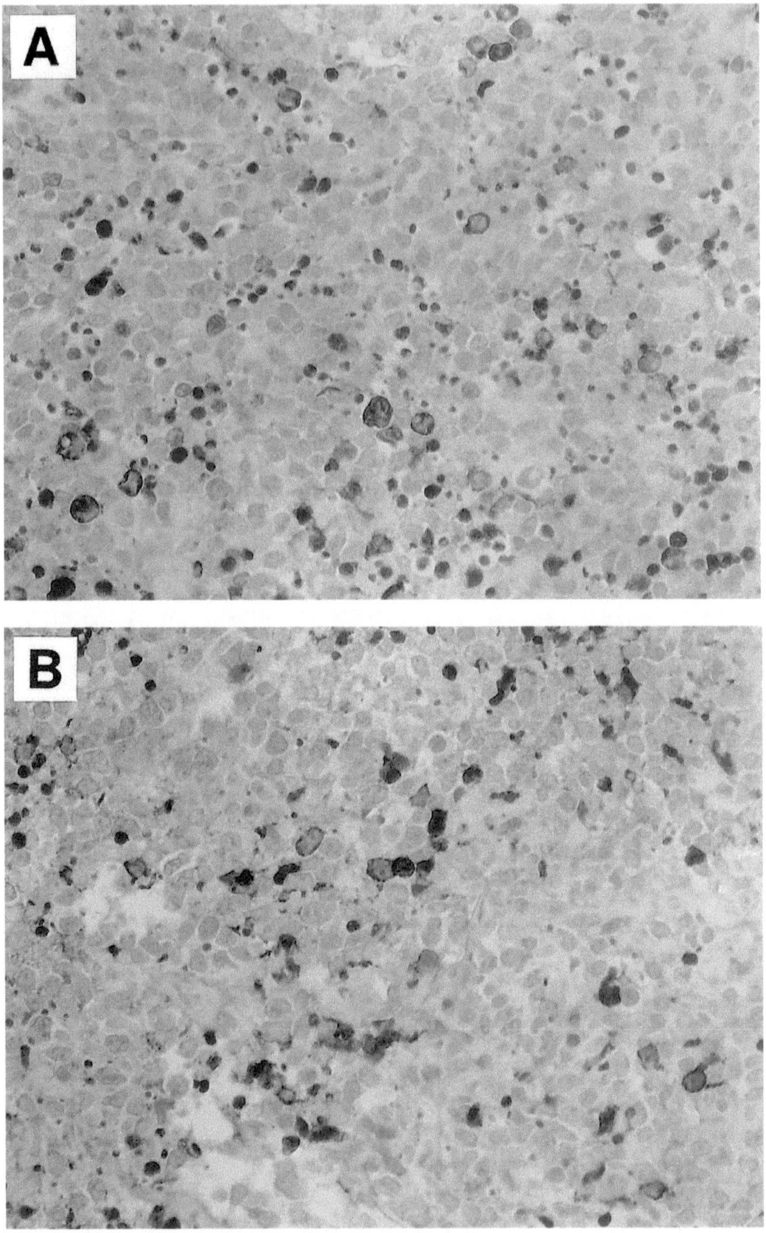

Fig. 3. Background with immunohistochemistry. The tissue is a lymph node in a patient with a history consistent with B-cell lymphoma. Both κ (**A**) and λ (**B**) were detected by immunohistochemistry; this presumably is background as the histology

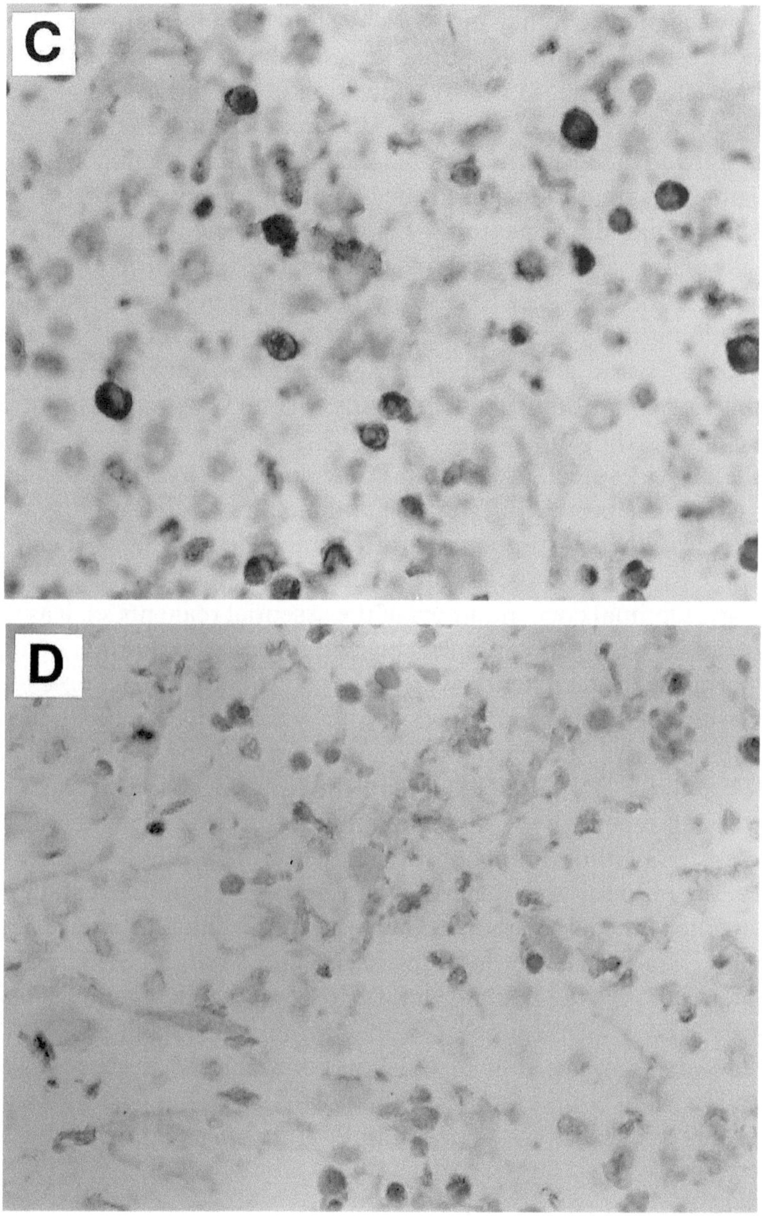

was consistent with a lymphoma, which should be monotypic. RT *in situ* PCR for κ mRNA showed a strong signal **(C)** and no signal for λ **(D)**, confirming that the immunohistochemistry signal was due to background.

is not to say that the technique of *in situ* hybridization has remained static. The detection threshold of this assay has improved substantially. Another point worth emphasizing about *in situ* hybridization is that one does not need to use radiolabeled probes (usually ^{35}S or ^{3}H) to maximize its sensitivity. Although true 10 yr ago, advances in nonisotopic labeling and, more importantly, detection systems has greatly enhanced the sensitivity when using such common labels as biotin and digoxigenin ***(7–12)***. Still, only the most aggressive salesman would claim (and incorrectly at that) that any given *in situ* system can routinely detect one DNA or RNA copy per cell. In my experience, this statement also applies to the newer generation of posthybridization "signal amplification systems" (such as the cascade amplification system) which are not able to routinely detect one copy per cell. **Figure 2** shows a comparison of different *in situ* methods and *in situ* PCR.

Despite the widespread use of both PCR and *in situ* hybridization in the last several years it has proved difficult to combine the two. The ability to accomplish PCR *in situ* hybridization in paraffin embedded tissue has been difficult for several reasons. One must expose the target DNA without destroying tissue morphology. Optimal concentrations of the essential reagents such as the primers, Mg^{2+}, and the DNA polymerase must be determined. Further, and perhaps most importantly, if the reaction were to be carried out directly on glass slides, loss of tissue adherence and tissue drying would have to be circumvented. These problems have been overcome to the point that one may reliably amplify both DNA and, for RNA, cDNA in paraffin-embedded, fixed tissues. A point worth stressing repeatedly is that background is a problem with ANY *in situ* methodology, including immunohistochemistry, *in situ* hybridization, and *in situ* PCR (**Fig. 3**) and, with under the appropriate conditions, can be successfully controlled in each of these assays. Stating it another way, background false-positive signals are not just a problem with *in situ* PCR. The purpose of this chapter is to provide readers with the protocols this laboratory has developed for the *in situ* localization of PCR-amplified DNA and cDNA. This will be preceded by a discussion of the key components of successful *in situ* hybridization as, of course, an in-depth knowledge of the mechanics of *in situ* hybridization is essential for performing PCR *in situ* hybridization.

2. Materials

1. Slide preparation: Use silane coated slides which may be purchased from Enzo Diagnostics (Farmingdale, NY)(*see* **Note 1**).
2. Fixative: Use 10% buffered formalin and fix the cells or tissue for 15–24 hours (*see* **Note 2**).
3. Protease digestion: Use 2 mg/mL of pepsin (Dako Corporation, Carpinteria, CA) at room temperature for 5–30 min (*see* **Note 3**).

4. Probe cocktail:
   Add together the following ingredients:
   a. 50 µL of deionized formamide (use 10 µL for oligoprobes and add 40 µL of water).
   b. 30 µL of 25% dextran sulfate.
   c. 10 µL of 20× saline sodium citrate (SSC).
   d. 10 µL of the probe (stock solution of 5–10 µg/mL) (*see* **Note 4**).
5. Posthybridization wash: The wash solution contains 2.5% bovine serum albumin (BSA) and 0.2× SSC (or 30 m*M* sodium chloride for full-length probes and 150 m*M* for oligoprobes) and is heated to 45°C (oligoprobes) or 60°C (for full-length probes) (*see* **Note 5**). The wash powder, which is rehydrated in 1 L of water, can be purchased from Enzo Diagnostics.
6. Detection systems:
   a. (For biotin system) Use a streptavidin–alkaline phosphatase (AP) conjugate (Enzo Diagnostics); for digoxigenin use antidigoxigenin–AP conjugate (1:200 dilution, Boehringer Mannheim, Indianapolis, IN)
   b. After using the AP conjugate, the slides should be placed in a solution of 0.1 *M* Tris-HCl (pH 9–9.5) and 0.1 *M* NaCl (detection reagent solution). The wash powder, which is rehydrated in 1 L of water, can be purchased from Enzo Diagnostics.
   c. The chromagens nitroblue tetrazolium and 5-bromo-4-chloro-3-indolylphosphate (NBT/BCIP) (Enzo Diagnostics) are added to the detection reagent. Use 50 µL each per 15 mL of the pH 9.5 buffer.
   d. Counterstain is nuclear fast red (Fisher Scientific, Pittsburgh, PA) (*see Note 6*).
7. The PCR step of PCR *in situ* hybridization:
   a. Use silane-coated glass slides.
   b. Fix the cells or tissue in 10% buffered formalin for 15–48 h.
   c. Deparaffinize tissue in xylene (5 min) and 100% ethanol (5 min).
   d. Digest in 2 mg/mL or pepsin or trypsin at room temperature for 5–30 min.
   e. The amplifying solution needs to contain 4.5 m*M* $MgCl_2$, 1 µ*M* primer, 2.5 U/25 µL of *Taq* polymerase (assuming BSA is added), 200 µ*M* of the dNTPs, and the GeneAmp kit buffer (Perkin Elmer) (*see* **Note 7**).

3. Methods

3.1. Running the Reaction

1. Place several 4-µm paraffin-embedded sections or two cytospins on a silane-coated glass slide.
2. Wash paraffin-embedded sections in xylene for 5 min; then in 100% ethanol for 5 min, then air dry (for tissue sections only).
3. Digest in pepsin; inactivate protease by washing in 0.1 *M* Tris-HCl, pH 7.5, and 0.1 *M* NaCl for 3 min.

4. Wash slides in 100% ethanol for 3 min, then air-dry.

For PCR step (if no PCR step, proceed to **step 11**)

5. Add 2.5 μL of GeneAmp buffer, 4.5 μL of $MgCl_2$ (25 m*M* stock), 4 μL of dNTPs (2.5 m*M* stock), 1 μL of primer 1 and primer 2 (each 20 μ*M* stock), and 11.2 μL of sterile water. Remove 4 μL (keep on ice for hot start). (Note: An alternative is to use a *Taq* polymerase such as AmpliGold (Perkin Elmer), which can be added at room temperature as the enzyme is not capable of DNA synthesis until a preheating step. One also has the option of using the Perkin Elmer AmpliClip and AmpliCover material with the AmpliGold, which obviates the need for the mineral oil overlay as well as the manual hot start maneuver).
6. Place solution on two separate sections; add plastic coverslip, anchor with nail polish.
7. Time delay file—keep at 82°C for 7 min.
8. At the onset of this file add 0.8 μL of *Taq* polymerase to the tube on ice.
9. At 75°C, lift one edge of the coverslip gently and add 2.4 μL to each section, overlay with preheated mineral oil.
10. Switch to time delay file—keep at 94°C for 3 min.
11. Link this time delay file to a cycling file of 55°C—2 min and 94°C—1 min for 20–40 cycles; at conclusion link to soak file of 4°C; remove mineral oil with xylene and ethanol washes, air dry.
12. Add 5–10 μL of the probe cocktail to a given tissue section
13. Overlay with plastic coverslip cut slightly larger than tissue section.
14. Place slide on hot plate at 95–100°C for 5 min.
15. Remove bubbles over tissue gently with a toothpick.
16. Place slides in humidity chamber at 37°C for 2 h.
17. Remove coverslips—hold down one end with fingernail and lift off coverslip with toothpick.
18. Place in wash solution for 10 min. at 60°C (assuming full-length probe).
19. Wipe off excess wash solution and put slides in a humidity chamber; do not let slides dry out.
20. Add appropriate alkaline phosphatase conjugate in humidity chamber.
21. Incubate for 30 min at 37°C.
22. Wash slides at room temperature for 3 min in a solution of 0.1 *M* Tris HCl, pH 9–9.5, and 0.1 *M* NaCl (detection reagent solution).
23. Place slides in detection reagent solution to which NBT/BCIP has been added.
24. Incubate slides for 30 min to 2 h, checking results periodically under microscope.
25. Counterstain with nuclear fast red and coverslip; view under the microscope (*see* **Notes 8–10**).

4. Notes

1. Perhaps the major technical advancement in the field dealt simply with the preparation of the glass slide. About 10 yr ago, it was routine to pretreat the slides with

materials such as poly-L-lysine, glue, or other adhesives to improve adherence (***13–16***). Although such pretreatments certainly worked better than untreated slides where most sections would fall off, in my experience the sections would, at best, remain on about 75% of the time. It should be noted adherence is much less a problem with cytologic preparations, such as Pap smears. One can perform *in situ* hybridization or *in situ* PCR using destained Pap smears on routine slides with excellent adherence (**Fig. 1**). Loss of adherence using tissue sections was circumvented by the use of organosilane, a chemical used in industry to treat glass. I have stored silane coated slides at room temperature in closed boxes and used these slides successfully as much as *10 yr* after pretreatment. If tissue sections fall off at a rate >5% and one is using commercially prepared silane slides, the problem is probably air bubbles under the tissue that usually reflects inexperience of the technician placing the sections on the slides.

2. Some groups recommend using frozen, unfixed tissues for *in situ* hybridization to avoid the protease digestion step. I do not recommend using frozen tissues for *in situ* hybridization. The morphology is at best poor, which defeats the purpose of the test. Further, some claim that protease digestion is still required with *in situ* hybridization with the use of frozen tissues so even this step cannot necessarily be omitted (***10,17***). There are many different fixatives that various laboratories use to process tissues. In my own experience, I have used buffered formalin, unbuffered formalin, Bouin's solution, and B5 (each of which contains picric acid), Zenker's solution (which contains mercury), and 95% ethanol. When using tissue sections, buffered formalin, pH 7.0, is the best fixative for *in situ* hybridization and PCR *in situ* hybridization, although other crosslinking fixatives such as glutaraldehyde and paraformaldehyde are acceptable. Fixatives that contain either picric acid or heavy metals may allow for successful *in situ* hybridization but this is dependent on the length of time of fixation (***15,18–22***). Two hours of fixation in Bouin's solution has a minimal effect on the intensity of the hybridization signal that may be completely eradicated after overnight (15 h) fixation; intermediate results are seen after 8 h of Bouin's fixation (***15,20***). However, fixation in a solution that contains a heavy metal or picric acid will not permit either PCR or PCR *in situ* hybridization (***19,23***).

   If only tissues fixed in either picric acid or a heavy metal are available along with frozen tissue, it is recommended that the frozen tissue be slowly thawed and then fixed in 10% buffered formalin overnight. Although some freezing artifact will be evident, both *in situ* hybridization and PCR *in situ* hybridization may be done with good results (G. J. Nuovo, *unpublished observations*).

   Cellular preparations, such as Pap smears and fine-needle aspirates, are usually fixed in an alcohol-based solution immediately after being obtained. One can use such specimens for successful *in situ* hybridization or PCR *in situ* hybridization. The key is to use a relatively weak protease digestion; I recommend proteinase K at 10 μg/mL of water for 10 min (**Fig. 1**). Further, if the cytology slide has been stained, then 15 min in acid alcohol (0.2 N HCl) is needed to remove the Pap stain, prior to performing the *in situ* hybridization or *in situ* PCR.

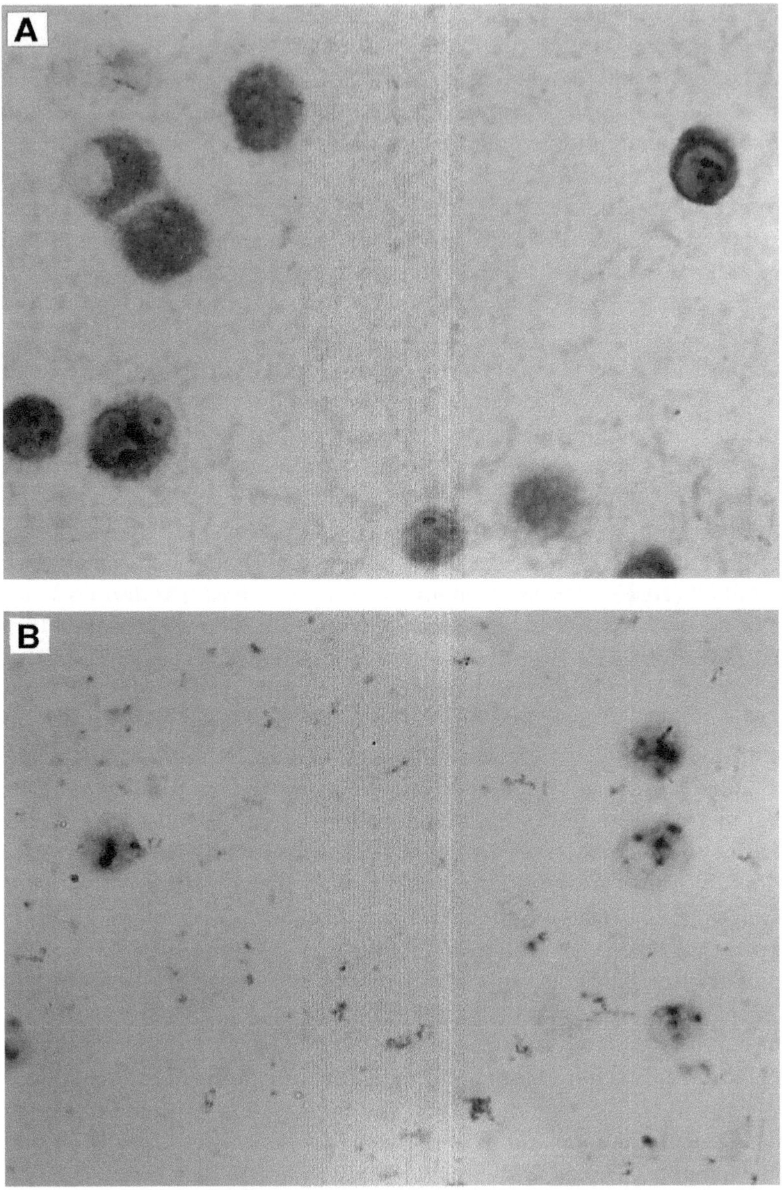

Fig. 4. Background with *in situ* hybridization—underprotease digestion. Caski cells, which contain about 500 copies of HPV 16 per cell, showed a weak signal and high background when the protease digestion was less than optimal (5 min of 2 mg/mL of pepsin—**A**). Note the intense, pinpoint signal and loss of background when the protease time was increased to 30 min **(B)**.

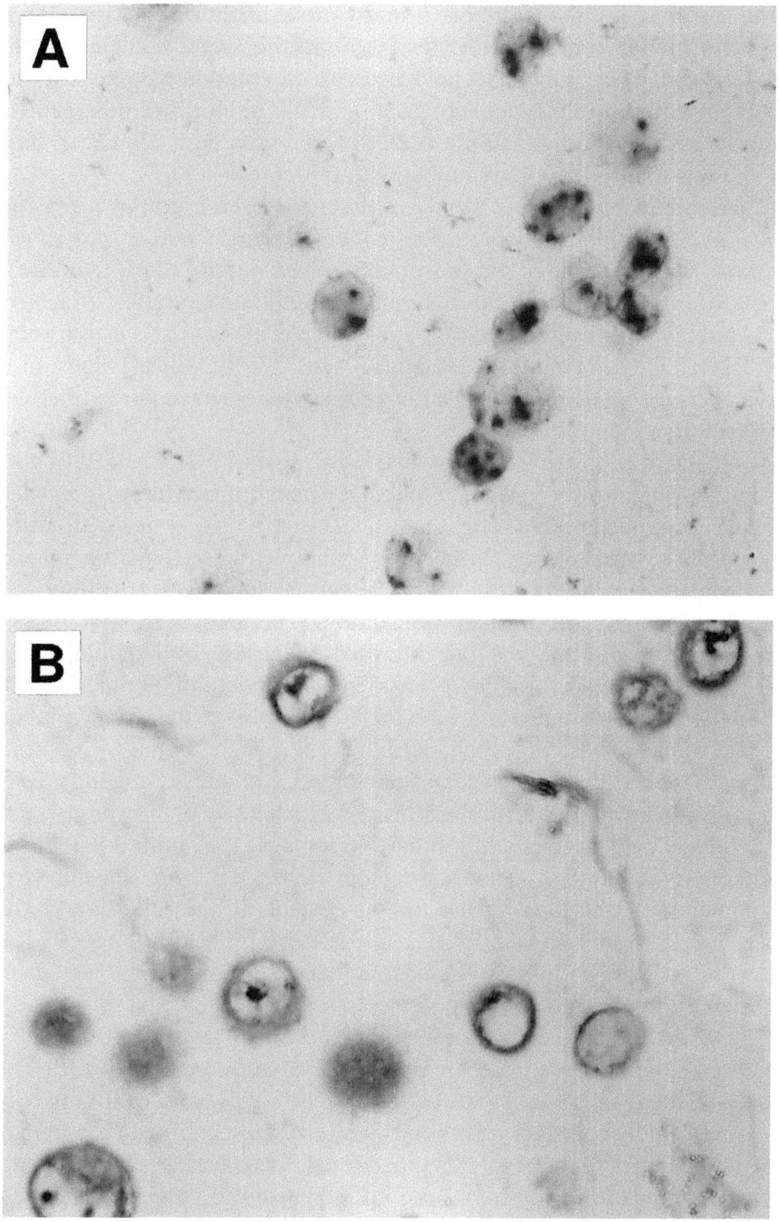

Fig. 5. Background with *in situ* hybridization—overprotease digestion. Note the similar intense, pinpoint signal in these Caski cells with optimal protease digestion (**A**). The signal is lost and the cell morphology is less clear when protease digestion is increased to 60 min (**B**); note the background evident as a cytoplasmic rim.

3. Fixatives, especially those whose primary mode of action is crosslinking such as formalin, hinder penetration of the probe to the target nucleic acid molecule ***(15,22–24)***. Different methods have been used to facilitate probe entry including treatment with various chemicals such as HCl, photofluor, detergents, heating (microwave), and sodium sulfite (***12***), to name but a few. However, most of the interest has focused on pretreatment with proteases.

   There are many different proteases that have been used for *in situ* hybridization and immunohistochemistry. These include trypsin, pepsin, proteinase K, and pronase. We have found that these are equivalent for *in situ* hybridization but each requires determining the optimal time and concentration which may vary considerably for different tissues. For example tissues such as biopsies of kidney, liver, and lymph nodes are relatively sensitive to protease digestion whereas cervical tissue and, especially, autopsy material often requires relatively more stringent digestion (***22***).

   Insufficient protease treatment is recognizable as a diminished or completely absent hybridization signal and, typically, a concomitant increase in background (**Fig. 4**). Too much protease treatment is easy to recognize as the tissue morphology will be destroyed (**Fig. 5**). It is important to emphasize that in my experience overprotease treatment is the *primary* cause of poor morphology for *in situ* hybridization and PCR *in situ* hybridization. If poor tissue morphology is a problem, then decrease either the time of protease digestion or the protease concentration 10-fold. Another variable to consider is the denaturation time. It has been my experience that a denaturation time of >5 min, or exposure of the glass slides to dry heat after protease digestion, can both diminish the signal with *in situ* hybridization as well as destroy tissue morphology (G. J. Nuovo, *unpublished observations*). To determine if the protease time was adequate, it is strongly recommend that one use one of several biotin labeled human DNA probes that are currently available (e.g., blur 8 probe which detects the repetitive *alu* sequence from Enzo Diagnostics). When the protease time is correct, all the cells should give an intense hybridization signal with the use of these probes; as suggested, many of these probes use repetitive *alu* sequences that are present in numbers much greater than 10,000 copies per cell (**Fig. 6**).

   The protease digestion time recommended above of 5–30 minutes, in my experience, will be adequate for about 95% of surgical biopsies. Brief mention should be made of what may be called "hyperfixed" samples. In my experience, these are usually cellular preparations which have been fixed for a long period of time (e.g., weeks) or a elevated temperatures. In such cases, I have needed to use up to 100 min of digestion in 2 mg/mL of pepsin to obtain the strong, optimal signal with *in situ* hybridization or *in situ* PCR. If this still does not work, then try proteinase K (1 mg/mL diluted in 2Á SSC) for 60 min; proteinase K has a much higher activity than pepsin or trypsin.
4. The function of the formamide and relatively low salt concentration is to facilitate denaturing of the probe and target DNA at 100°C, about 40°C above the melting temperature (T_m) of homologous hybridized DNA for hybrids of 100

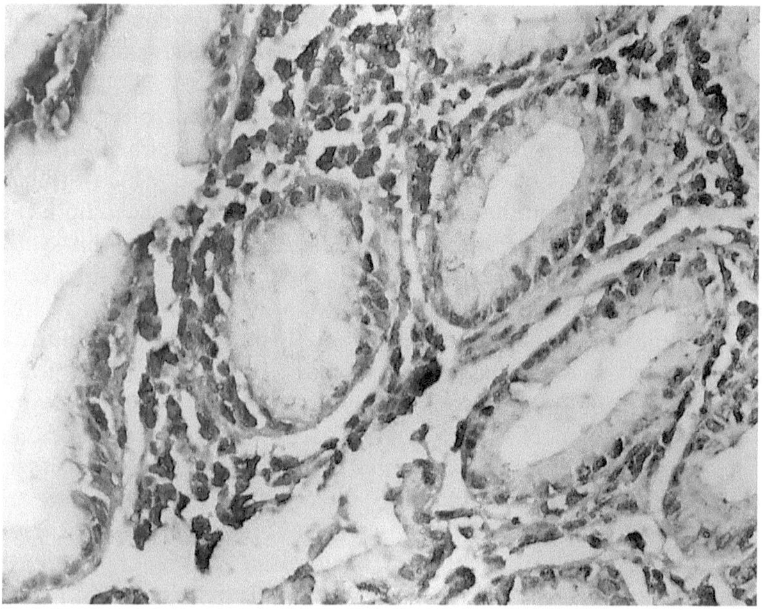

Fig. 6. Detection of the *alu* repetitive sequence with the Blur 8 probe. Note how all cell types show an intense signal with the Blur 8 probe (Enzo Diagnostics), indicating that the protease, hybridization, and detection systems were functionally correct.

basepairs (bp), as is typical for so-called full-length probes (probes made from templates at least 200 bp in size). Bromley et al. have done extensive work correlating the concentration of the probe and the intensity of the hybridization signal under a wide variety of conditions (*25*). The probe concentration we typically use is 1 µg/mL. This amount is rarely associated with background problems for the nonisotopic labeled probes. However if background is a problem the concentration of the probe may be decreased 10-fold, which—if background is reduced—invariably will lead to a concomitant increase in the intensity of the hybridization signal (*see* 8) (*25*).

5. Background is easily corrected with the most common nonisotopic systems—biotin and digoxigenin (**Fig. 4**). The conditions listed earlier usually readily disallow the relatively few hydrogen bonds between the large (100–200 bp) probe and nontarget molecules such as cellular proteins but still maintain the probe–target complex. However, it is important to emphasize that for oligoprobes (20–40 bp), one needs to use much less stringent conditions (150 m*M* salt and 45°C for 10 min), or risk losing the entire signal. It should also be stressed that the window of high signal to low background is very narrow with oligoprobes and is dependent on not just the stringent conditions but also on the concentration of the probe and time of hybridization. I prefer to keep oligoprobe concentration

and hybridization time constant and vary the stringency conditions, but one can adjust any of these three variables to try to maximize signal and reduce background.

6. After the posthybridization wash one is left with a target–probe complex. Labeled nucleotides are incorporated into the probe. A key component of the biotin or digoxigenin systems is the enzyme alkaline phosphatase, which will be attached to the probe–target complex. For biotin this is readily accomplished with a streptavidin–alkaline phosphatase conjugate. An advantage of this system is that any immunohistochemistry laboratory will have extensive experience with such conjugates and thus be aware of the nuances of its use. For the digoxigenin system one employs an antibody against digoxigenin which is conjugated to the alkaline phosphatase. Although a wide variety of chromagens are available, I recommend 5-bromo-4-chloro-3-indolylphosphate which in the presence of nitroblue tetrazolium (NBT/BCIP) yields a blue precipitate. The counterstain, which stains the cytoplasm and negative nuclei pale pink and is nuclear fast red.
7. Note that the optimal concentrations of the Mg^{2+} and *taq* polymerase are greater than those for standard PCR. This may reflect difficulty in entry of these reagents to the site of DNA amplification and, in part, sequestration of the Mg^{2+} by cellular components. Consistent with this hypothesis is our observation that one may use 10-fold less *taq* polymerase with the addition of 1 mg/mL of BSA to the amplifying solution in PCR *in situ* hybridization; BSA can block absorption of the enzyme to the glass slide or plastic coverslip (***23***).
8. The most common problems encountered with *in situ* hybridization are background and the absence of a signal. Background may be defined as the presence of a hybridization signal with a specific probe in areas of the tissue where the signal should not be present (i.e., normal endocervical cells or in basal cells with HPV). Of course, in some instances one may not be sure where the *in situ* signal should localize. A more strict definition of background would be a hybridization signal in tissues or cells known not to contain the target of interest (this may be determined by solution phase PCR or, for some viral infections, the lack of the diagnostic histologic changes). Background is the result of nonspecific binding of the probe to nontarget molecules, including cellular proteins (cytoplasmic) and nucleic acids (nuclear). Two simple and logical ways to deal with background are to decrease the concentration of the probe and/or to increase the stringency of the posthybridization wash. If background is a problem, first try decreasing the concentration of the probe 10-fold. If background persists, decrease the concentration of the sodium in the post hybridization wash 10-fold.

   The most obvious potential problem with *in situ* hybridization is the absence of a hybridization signal. I recommend following a flow-chart type of problem solving tree, which is presented in **Fig. 7**.
9. The probe size for standard *in situ* hybridization is 100–250 bp in size. However, one may want to use much smaller (20–40 bp) probes called oligoprobes. There are two main reasons to use oligoprobes: they are more readily available than the larger probes which require a cloned sequence of DNA; one only needs to know

FLOW CHART FOR LACK OF SIGNAL
WITH IN SITU HYBRIDIZATION

Use POSITIVE CONTROL DNA (or RNA) probe

NO SIGNAL	SIGNAL
1. Is protease time correct?	
A. Tissue morphology poor – OVERDIGESTION (repeat with less protease) B. Tissue morphology good – UNDERDIGESTION (repeat with more protease, may need to use 1 hr)	1. Check histology of tissue (is there any histologic evidence of target) (do PCR to determine if target is present in tissue) 2. Is probe correctly labeled? (Check incorporation of labeled nucleotide in probe with filter)

IF STILL NO SIGNAL PROBLEM MAY BE (in order of likelihood)

2. Alkaline phosphatase conjugate no longer active (more likely with anti-digoxigenin conjugate)
3. NBT/BCIP chromogen no longer active
4. Fixative either contains picric acid or heavy metal (eg Bouin's, Zenker's)

Fig. 7. Flow chart for a negative result with the *in situ* hybridization. The figure details a step-by-step approach to follow if a hybridization signal is not evident with *in situ* analysis.

the sequence of the target of interest, readily available in the literature, to generate an oligoprobe. Second, one is obliged to use an oligoprobe internal to the sequence being amplified in solution phase PCR in order to assure themselves that the signal is indeed the PCR product (as opposed to a signal due to primer oligomerization). Because oligoprobes are much shorter (20–40 bp vs 100–250 bp) than a "standard" probe, there is a substantial reduction in the number of basepair matches and thus the strength of the hybridized complex compared to the larger probes. The practical consequence is that the wash conditions must be carefully chosen so as to minimize background but not to lose the signal. In practical terms, I have seen the signal lost for a 20-mer oligoprobe with a posthybridization wash in 30 m*M* salt at 45°C; under these conditions the signal for a larger homologous probe would remain intact. Hence, I use different probe cocktail and posthybridization wash conditions for oligoprobes as listed previously.

It should be stressed that primer oligomerization does NOT appear to occur inside nuclei during *in situ* PCR *(33)*. This may reflect the relatively high protein concentration inside the nucleus; single stranded binding proteins can inhibit primer oligomerization during solution phase PCR *(33)*. This observation has important practical implications for PCR *in situ* hybridization. Specifically, one may use the full-length probe with PCR *in situ* hybridization, even though it includes the region of the primer, and still detect target specific signal assuming that other potential causes of background have been eliminated. Full-length probes permit much more stringent washes due to their wide range of signal to background. Thus, it is much easier to eliminate background and still preserve signal with a full-length probe relative to an oligoprobe.

10. Direct incorporation of reporter molecules is possible for DNA targets with PCR *in situ* but only under strictly defined conditions. The inclusion of a labeled nucleotide in the amplifying solution is the major modification in this technique compared to PCR *in situ* hybridization, where the PCR product is not directly labeled but rather detected with a labeled probe. Most of our work with *in situ* PCR has focused on the labeled nucleotide digoxigenin dUTP (dig dUTP–Enzo Diagnostics) ***(26–30)***. There are only two modifications to the protocol listed for PCR *in situ* hybridization. First, 10 µ*M* dig dUTP is added to the amplifying mixture. Second, after completing the amplifying reaction and removing of the coverslip and mineral oil, the slides are washed for 10 min at 65°C in a solution of 2% BSA and 0.1× SSC to remove unincorporated dig dUTP and, more importantly, primer oligomers that may have formed in the overlying amplifying solution (as compared to inside the cell). After this wash, the digoxigenin that has been incorporated into the amplified DNA may be detected according to the manufacturer's protocol (Boerhinger Mannheim; I use a 1:200 dilution of the antibody). It is important to note, as stressed later, that one must use the hot start maneuver to perform target specific incorporation of the labeled nucleotide during *in situ* PCR for DNA targets. Further, one must use cells or tissue that has NOT been exposed to dry heat prior to *in situ* PCR, which induces a primer independent signal due to DNA nicking ***(33)***. This negates the use of *in situ* PCR for DNA targets with paraffin-embedded tissue, which is heated at 65°C for 4 h prior to embedding. For paraffin-embedded tissues, one must use PCR *in situ* hybridization with a probe step after the PCR. The DNase digestion step that is done after protease digestion for reverse transcription (RT) *in situ* PCR eliminates all the DNA synthesis pathways (mispriming, target-specific and primer independent DNA repair). This allows for the target-specific incorporation of the labeled nucleotide in the amplified cDNA ***(33)***.

With regards to specificity, in solution phase PCR, two pathways compete with target-specific DNA synthesis. These are mispriming and primer oligomerization ***(30)***. If the hot start modification is not employed, the mispriming and primer oligomerization pathways can easily overwhelm target specific DNA synthesis such that a large amount of DNA is synthesized but it is MOSTLY NONSPECIFIC! This is not surprising when one considers that there is far more nontarget and primer DNA in a reaction mixture relative to target DNA. It has been shown that under non hot start conditions that the detection threshold for the target of interest with solution phase PCR may be greater than several thousand copies per 100,000 cells, not the 1–100 copies most articles quote ***(30,31)***. However, nonspecific DNA synthesis is greatly curtailed by the hot start modification. The end result is that one can reliably detect 10 copies per tissue with the hot start modification of PCR ***(30,31)***. Similarly, the hot start modification of PCR *in situ* hybridization (or with direct labeling using *in situ* PCR), by inhibiting mispriming, allows for the reproducible detection of one target copy per cell using a single primer pair ***(30,33)***.

Two different approaches may be used to demonstrate the specificity of hot start *in situ* PCR. First, different cell populations can be mixed and a primer pair is used that is able to amplify a target in only one of the two populations. Second, one may use "irrelevant primers," primer pairs that could not possibly find targets in the cells being studied.

SiHa cells are from a human cervical cancer cell line which, as expected, contain HPV DNA although only one viral copy per cell. These were mixed with peripheral blood leukocytes (PBMs) from a noninfected individual, although HPV cannot infect PBMs had an HPV infection been present at some other site. Standard conditions refer to experiments in which all reagents were added BEFORE raising the temperature of the heating block. Whereas SiHa cells might give specific or nonspecific product, amplified DNA from PBMs must be nonspecific when the HPV specific primers are employed. The digoxigenin would become incorporated into both target-specific and nonspecific amplified DNA. Under hot start conditions with a single HPV 16 primer pair only some of the cells incorporated digoxigenin. The negative cells proved to be the leukocytes as they reacted with an antibody against leukocyte common antigen in a double labeling technique. Under standard conditions all of the cells, including the leukocytes, incorporated digoxigenin (*see* **Fig. 2**, **ref.** ***30***). These results provided reassurance that the hot start modification greatly inhibited nonspecific pathways ***(30)***.

In an analogous experiment, measles infected HeLa cells were mixed with the PBMs of a noninfected individual. The measles infected HeLa cells are multinucleated and thus easily differentiated cytologically from the single nucleated leukocytes. These mixing experiments proved that only the measles-infected HeLa cells had a detectable signal with direct incorporation and measles specific primers if the hot start modification was employed with the RT *in situ* PCR technique (see **refs.** ***32, 33***). If the hot start modification was omitted then both the HeLa cells and PBMs had detectable signal.

We typically employ irrelevant primers as one of our standard negative controls with PCR *in situ* hybridization. For HPV experiments, we use primers specific for either rabies or hantavirus as, obviously, neither should be present in routine surgical biopsies. It is important to stress that one should use the same (i.e., HPV specific) probe under these conditions as it is our experience that the probe is the primary cause of background with *in situ* hybridization or *in situ* PCR. It follows that a negative control that uses a different probe (e.g., a plasmid vector) will not be very helpful in determining whether the signal evident with the HPV probe is target specific of not.

5. Concluding Comments

1. Brief mention is made of our RT PCR *in situ* hybridization protocol and results. More detailed information is available ***(32–36)***. The model system described is with the RNA virus hepatitis C (and other hepatotrophic viruses such as hepatitis

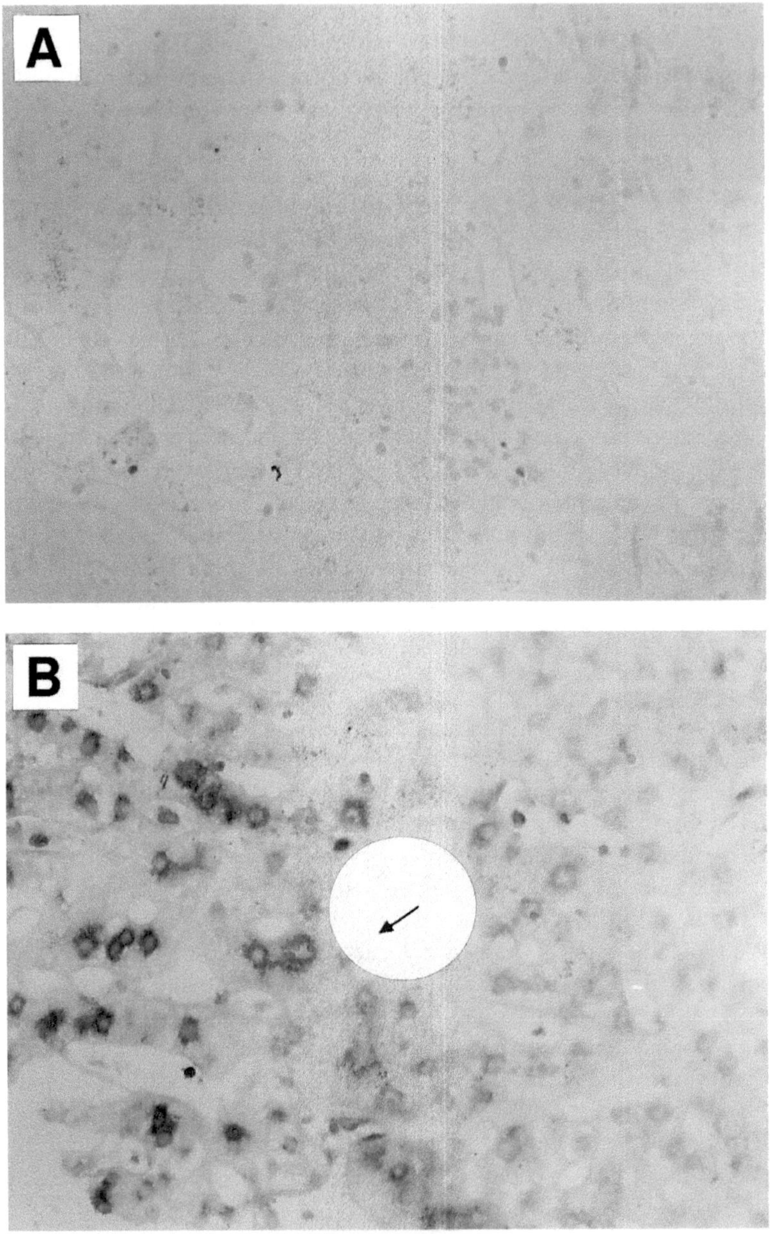

Fig. 8. Detection of hepatitis G RNA with RT *in situ* PCR. The liver biopsy is from a patient who had a liver transplant for hepatitis C infection, and subsequently demonstrated persistent chronic hepatitis of unknown etiology. Hepatitis C RNA was not detected by RT *in situ* PCR (**A**). However, hepatitis G RNA was detected by this methodology (**B** and, at higher magnification, **C**); note the cytoplasmic/perinuclear based

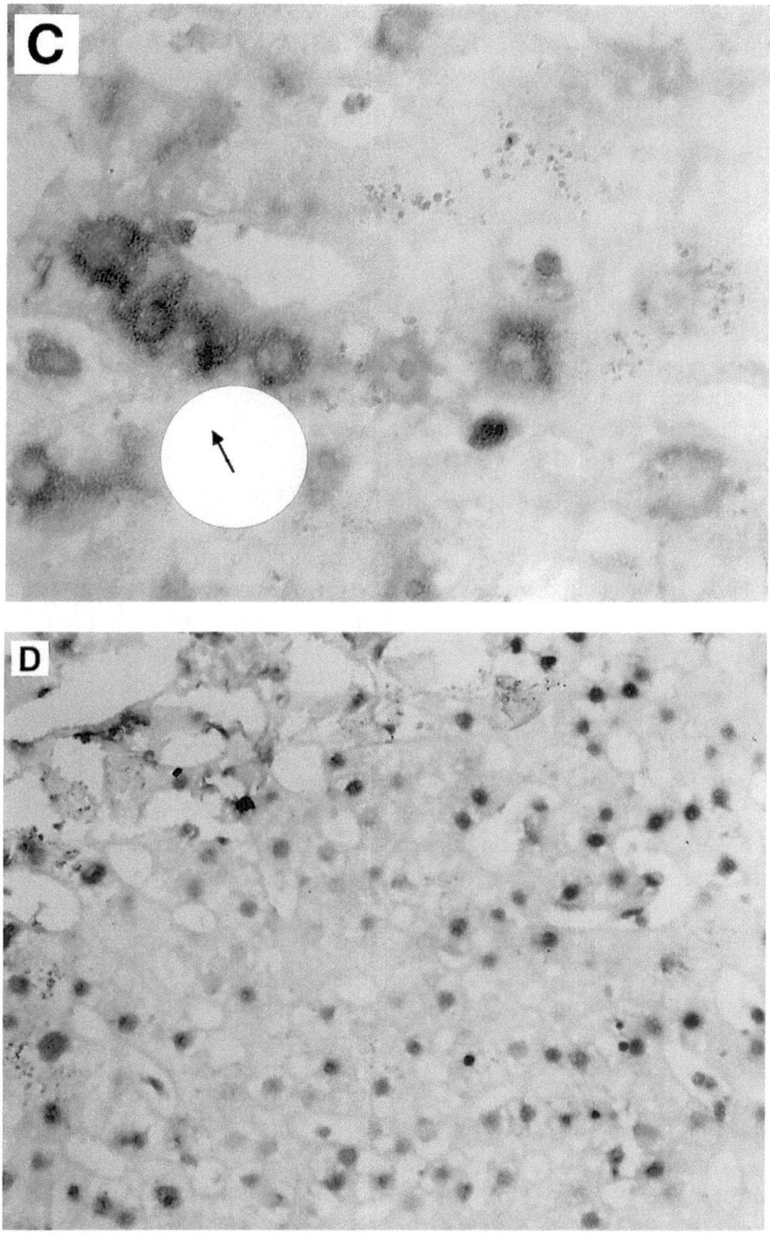

signal evident in hepatocytes, which are the target cell of this virus. Compare this distribution pattern with the serial section where DNase digestion was not done (**D**) where an intense nuclear signal, due primarily to DNA repair, is seen. This highlights the need for optimal protease digestion and DNase digestion when doing RT *in situ* PCR.

E and G). All of these reactions were performed directly on glass slides. The key point to emphasize about our protocol is the overnight pretreatment in RNase-free DNase. First, it is important to realize that one must use long-term digestion. The end point of determining the proper DNase digestion time is: *the inability to synthesize genomic based DNA with either target specific or irrelevant primers or via the primer independent pathway*. In our experience, at least 7 h of DNase digestion AFTER optimal protease digestion is needed to eliminate all *in situ* nonspecific DNA synthesis.

The inability to synthesize DNA in the liver tissue after DNase digestion permits target specific incorporation of the labeled nucleotide into the cDNA made during the RT step of the RT *in situ* PCR reaction (**Fig. 8**). Also note that one may exploit the fact that there is invariably DNA synthesis in the non-DNase-treated paraffin-embedded tissue section by using it as a positive control for the PCR and subsequent dig dUTP detection steps and the DNase non-RT section as the negative control (**Fig. 8**).

In the RT step, the solution is covered with a plastic coverslip, anchored with nail polish and overlaid with mineral oil as described and illustrated for PCR *in situ* hybridization. We use the same reagent concentrations for RT as listed in the RT PCR kit (Perkin Elmer, Norwalk, CT). Another important advantage to remember is to use the polymerase rTth. This allows one to do the RT and PCR steps under the same amplifying solution ***(33–36)***; this one step protocol simplifies the procedure.

Amplified hepatitis C, E, and G cDNA were detectable in several of the liver biopsies, all from patients with histologic evidence of the infection either as chronic hepatitis or cirrhosis. The specificity was demonstrated by omitting the direct incorporation of dig d-UTP in the PCR step and using a labeled internal oligoprobe and demonstrating in serial sections that the same cells were positive. Note again the importance of the essential negative and positive controls. A common cause of a false-positive result with RT *in situ* PCR is the persistence of genomic-based DNA synthesis owing to inadequate protease digestion which does not permit sufficient access of the native DNA to the DNase. This will be recognized by a nuclear-based signal in the negative control and can be rectified by increasing the time of protease digestion.

2. *In situ* hybridization and *in situ* PCR have added greatly in the diagnosis of HPV infections (and just as importantly, ruling out the infection) in clinical samples and in the understanding of oncogenesis as defined by cervical cancer. A common clinical problem that faces both the gynecologist and pathologist are genital tract biopsies suggestive of but not diagnostic of SIL. The absence of HPV by either *in situ* hybridization or *in situ* PCR, in conjunction with the equivocal histologic and cytologic findings, in my opinion RULES out SIL as the diagnosis and demonstrates that these are nonspecific changes. To state it in other words, this shows that the patient does NOT have a sexually transmitted disease by an oncogenic virus, that would require additional therapy, such as surgical excision or laser treatment.

With respect to the evolution of cervical cancer, it has been well documented that certain HPV genes (called open reading frames or ORFs) can dictate the synthesis of proteins that can transform normal cells. Indeed, ORF E6 and E7 have been shown to promote cell growth and transformation by the competitive inhibition of the important anti-oncogenes *Rb* and *p53*, respectively. Another anti-oncogene, *p16*, has received a lot of attention recently as being important in the evolution of squamous cell cancers in general; most cervical cancers are of this cell type. *p16* is inactivated by the hypermethylation of its promoter ***(37)***. This can be detected by using a process called methylation-specific solution-phase PCR. We have adapted the technique to an *in situ* PCR assay. With respect to invasion of the underlying stroma, certain cellular enzymes, including the matrix metalloproteases (MMPs), especially MMP-92 and MMP-72, are known to play a role in this key process. Their inhibitors, the tissue inhibitors of metalloproteases (TIMPs), can reduce the aggressiveness of invasive cell lines using in vitro systems. Detailed analysis of cervical precancers and cancers for HPV E6, E7, *p16* inactivation, and MMP plus TIMP expression using *in situ* PCR methods have shown the following: HPV E6 and E7 are present in about 25% of cervical SILs and invasive cancers, and their presence does not correlate with depth of invasion; *p16* is inactivated in a varying number of low-grade SIL cells, but is invariably inactivated in high-grade cervical SILs and invasive cancers; MMP and TIMP expression is rarely detected in high-grade cervical SILs, but is commonly found in invasive cancer of the cervix ***(36–38)***. Increased invasion was associated with a marked increase in the percentage of cells that contained either MMP-92 or MMP-72 and a significant decrease in the cells with TIMP-1 or TIMP-2 as determined by RT *in situ* PCR. These data highlight the utility of the high sensitivity and cell localizing ability of *in situ* amplification systems for the study of the pathogenesis of HPV-related infections of the cervix and lower genital tract.

Acknowledgments

The authors greatly appreciate the technical and material assistance of Drs. John Atwood, Will Bloch, Deborah French, Dennis Groff, Larry Haff, Brian Holaway, John Sninsky, and Dr. Selaki. Ms. Angella Forde, Frances Gallery, and Kim Rhatigan provided expert technical assistance. Ms. Phyllis MacConnell has been especially helpful with the technical aspects of the *in situ* PCR procedures. We are especially indebted Mr. S. B. Lewis for his financial support. This work was supported by a grant from the Lewis Foundation to GJN.

References

1. deVilliers, E. M., Schneider, A., and Miklaw, H.(1987) Human papillomavirus infections in women with and without abnormal cervical cytology. *Lancet* **i,** 703–706.

2. Nuovo, G. J. and Cottral, S. (1989) Occult infection of the uterine cervix by human papillomavirus in postmenopausal women. *Am. J. Obstet. Gynecol.* **160,** 340–344.
3. Nuovo, G. J., Darfler, M. M., Impraim, C. C., and Bromley, S. E. (1991) Occurrence of multiple types of human papillomavirus in genital tract lesions: analysis by in situ hybridization and the polymerase chain reaction. *Am. J. Pathol.* **58,** 518–523.
4. Nuovo, G. J., Hochman, H., Eliezri, Y. D., Comite, S., Lastarria, D., and Silvers, D. N. (1990) Detection of human papillomavirus DNA in penile lesions histologically negative for condylomata: analysis by in situ hybridization and the polymerase chain reaction. *Am. J. Surg. Pathol.* **14,** 829–836.
5. Nuovo, G. J. (1990) Human papillomavirus DNA in genital tract lesions histologically negative for condylomata, analysis by in situ, Southern blot hybridization and the polymerase chain reaction. *Am. J. Surg. Pathol.* **14,** 643–651.
6. Nuovo, G. J. (1989) A comparison of different methodologies (biotin based and 35S based) for the detection of human papillomavirus DNA. *Lab. Invest.* **61,** 471–476.
7. Crum, C. P., Nuovo, G. J., Friedman, D., and Silverstein, S. J. (1988) A comparison of biotin and isotope labeled ribonucleic acid probes for in situ detection of HPV 16 ribonucleic acid in genital precancers. *Lab. Invest.* **58,** 354–359.
8. Nuovo, G. J. (1989) A comparison of slot blot, Southern blot and in situ hybridization analyses for human papillomavirus DNA in genital tract lesions. *Obstet. Gynecol.* **74,** 673–677.
9. Ostrow, R. S., Manias, D. A., Clark, B. A., Okagaki. T., Twiggs, L. B., and Faras, A. J. (1987) Detection of human papillomavirus DNA in invasive carcinomas of the cervix by in situ hybridization. *Cancer Res.* **47,** 649–653.
10. Walboomers, J. M. M., Melchers, W. J. G., and Mullink, H. (1988) Sensitivity of in situ detection with biotinylated probes of human papillomavirus type 16 DNA in frozen tissue sections of squamous cell carcinoma of the cervix. *Am. J. Pathol.* **131,** 587–594.
11. Crum, C. P., Symbula, M., and Ward, B. E. (1989) Topography of early HPV 16 transcription in high-grade genital precancers. *Am. J. Pathol.* **134,** 1183–1188.
12. Nagai, N., Nuovo, G. J., Friedman, D., and Crum, C. P. (1987) Detection of papillomavirus nucleic acids in genital precancers with the in situ hybridization technique, a review. *Int. J. Gynecol. Pathol.* **6,** 366–379.
13. Crum, C. P., Nuovo, G., Friedman, D., and Silverstein, S. J. (1988) Accumulation of RNA homologous to human papillomavirus type 16 open reading frames in genital precancers. *J. Virol.* **62,** 84–90.
14. Crum, C. P., Nagai, N., Levine, R. U., and Silverstein, S. J. (1986) In situ hybridization analysis of human papillomavirus 16 DNA sequences in early cervical neoplasia. *Am. J. Pathol.* **123,** 174–182.
15. Nuovo, G. J. and Silverstein, S. J. (1988) Comparison of formalin, buffered formalin, and Bouin's fixation on the detection of human papillomavirus DNA from genital lesions. *Lab. Invest.* **59,** 720–724.

16. Nuovo, G. J., Friedman, D., Silverstein, S. J., and Crum, C. P.(1987) Transcription of human papillomavirus type 16 in genital precancers. *Cancer Cells* **5,** 337–343.
17. Lebargy, F., Bulle, F., Siegrist, S., Guellaen, G., and Bernaudin, J. (1990) Localization by in situ hybridization of γ glutamyl transpeptidase mRNA in the rat kidney using 35S-labeled RNA probes. *Lab. Invest.* **62,** 731–735.
18. McAllister, H. A. and Rock, D. L. (1985) Comparative usefulness of tissue fixatives for in situ viral nucleic acid hybridization. *J. Histochem. Cytochem.* **33,** 1026–1032.
19. Greer, C. E., Peterson, S. L., Kiviat, N. B., and Manos, M. M. (1991) PCR amplification from paraffin-embedded tissues: effects of fixative and fixative times. *Am. J. Clin. Pathol.* **95,** 117–124.
20. Nuovo, G. J. (1989) Buffered formalin is the superior fixative for the detection of human papillomavirus DNA by in situ hybridization analysis. *Am. J. Pathol.* **134,** 837–842.
21. Tournier, I., Bernuau, D., Poliard, A., Schoevaert, D., and Feldmann, G. (1987) Detection of albumin mRNAs in rat liver by in situ hybridization, usefulness of paraffin embedding and comparison of various fixation procedures. *J. Histochem. Cytochem.* **35,** 453–459.
22. Nuovo, G. J. (1991) Comparison of Bouin's solution and buffered formalin fixation on the detection rate by in situ hybridization of human papillomavirus DNA in genital tract lesions. *J. Histotech.* **14,** 13–18.
23. Nuovo, G. J., Gallery, F., Hom, R., MacConnell, P., and Bloch, W. (1993) Importance of different variables for optimizing in situ detection of PCR-amplified DNA. *PCR Method Applic.* **2,** 305-312.
24. Goelz, S. E., Hamilton, S. R., and Vogelstein, B. (1985) Purification of DNA from formaldehyde fixed and paraffin embedded tissue. *Biochem. Biophys. Res. Commun.* **130,** 118-124
25. Bromley, S. E., Darfler, M. M., Hammer, M. L., Jones-Trower, A., Primus, M. A., and Kreider, J. W. (1990) In situ hybridization to human papillomavirus DNA in fixed tissue samples, Comparison of detection methods. In *Papillomaviruses,* Zurhausen, H. and Broker, T., eds., Wiley-Liss, New York.
26. Nuovo, G. J., Margiotta, M., MacConnell, P., and Becker, J. (1992) Rapid in situ detection of PCR-amplified HIV-1 DNA. *Diagn. Mol. Pathol.* **1,** 98–102.
27. Nuovo, G. J., Gallery, F., and MacConnell, P. (1992) Analysis of the distribution pattern of PCR-amplified HPV 6 DNA in vulvar warts by in situ hybridization. *Mod. Pathol.* **5,** 444–448.
28. Nuovo, G. J., Becker, J., MacConnell, P., Margiotta, M., Comite, S., and Hochman, H. (1992) Histological distribution of PCR-amplified HPV 6 and 11 DNA in penile lesions. *Am. J. Surg. Pathol.* **16,** 269–275.
29. Nuovo, G. J., MacConnell, P., Forde, A., and Delvenne, P. (1991) Detection of human papillomavirus DNA in formalin fixed tissues by in situ hybridization after amplification by PCR. *Am. J. Pathol.* **139,** 847–854.
30. Nuovo, G. J., Gallery, F., MacConnell, P., Becker, J., and Bloch, W. (1991) An improved technique for the detection of DNA by in situ hybridization after PCR-amplification. *Am. J. Pathol.* **139,** 1239–1244.

31. Chou, Q., Russell, M., Birch, D. E., Raymond, J., and Bloch, W. (1992) Prevention of pre-PCR mis-priming and primer dimerization improves low-copy-number amplifications. *Nucleic Acids Res.* **20,** 1717–1723.
32. Nuovo, G. J., Gorgone, G., MacConnell, P., and Goravic, P. (1992) In situ localization of human and viral cDNAs after PCR-amplification. *PCR Method Applic.* **2,** 117–123.
33. Nuovo, G. J. (1996) PCR In Situ Hybridization, Protocols and Applications, 3rd edit., Lippinicott-Raven, New York.
34. Nuovo, G. J. and Alfieri, M. L. (1996) AIDS dementia is associated with massive, activated HIV-1 infection and concomitant expression of several cytokines. *Mol. Med.* **2,** 358–366.
35. Nuovo, G. J. (1997) In situ detection of PCR-amplified metalloprotease cDNAs, their inhibitors, and human papillomavirus transcripts in cervical carcinoma cell lines. *Int. J. Cancer* **71,** 1056–1060.
36. Nuovo, G. J. (1998) Histologic distribution of hepatitis A,B,C,D,E and G with the concomitant cytokine response in liver tissues. *Diag. Mol. Pathol.* **7**, 267–275.
37. Nuovo. G. J., Baylin, S. B., Bolinsky, S., Herman, J. G., and Plaia, T. (1999) In situ detection of the hypermethylation-induced inactivation of the p16 gene as an early event in oncogenesis. *Proc Natl Acad Sci USA* **96,** 12754–12759.
38. Nuovo, G. J. (1998) Detection of human papillomavirus DNA in Papanicolaou smears: correlation with pathologic and clinical findings. *Diag. Mol. Pathol.* **7,** 158–163.

13

Quantitation of *In Situ* Hybridization Using Image Analysis of Radioactively Labeled RNA Probes

Teresa Bisucci, Tim D. Hewitson, and Ian A. Darby

1. Introduction

Quantitative analysis of messenger RNA (mRNA) from tissue homogenates, cell extracts, or fixed tissue sections is vital for studies involving gene regulation and expression. Quantitative analysis of mRNA allows the investigator to establish the transcription level of particular genes either in relative or absolute terms. Traditionally Northern blot analysis is a comparative technique that detects the amount of mRNA of the gene of interest normalized to the amount of a housekeeping gene ***(1,2)***. This is generally performed by densitometry of band intensities from an autoradiograph. The method has been employed for more than two decades, and its relative simplicity has made it the first port of call when examining RNA expression. It is particularly useful for examining mRNA transcription in tissue/cells exposed to various treatments. While Northern analysis will inform the investigator of the size and relative abundance of the mRNA of interest, it is a limited technique. Some of the limitations include: inefficient transfer of the RNA to the filter, its relative insensitivity for examining smaller quantities of mRNA and the saturable nature of autoradiography when film is used ***(3)***. The last limitation has been overcome by phosphorimaging, which allows measurement of band intensity in a linear fashion. Therefore, Northern blotting is best used as a semiquantitative method of examining relatively abundant mRNA.

A similar yet simplified version of Northern analysis is called dot blot analysis, the only exception is that the RNA is not fractionated through a denaturing gel ***(4)***. The RNA is directly placed on the filter, immobilized, and probed usually with a radioactively labeled probe. After hybridization with the radio-

From: *Methods in Molecular Biology, vol. 179: Gene Probes: Principles and Protocols*
Edited by: M. Aquino de Muro and R. Rapley © Humana Press Inc., Totowa, NJ

active probe the filter is exposed to an autoradiographic film and once again the signals on the autoradiograph may be analyzed using densitometry. This technique is more limited than Northern analysis because the RNA is not fractionated through a gel and therefore it is impossible to determine if the hybridization has been specific and no information is provided as to the size of the mRNA species on the blot. Dot blots have become less frequently used as other techniques have become available.

Hybridization of the radioactive probe to the target mRNA can also be performed in solution, this is termed ribonuclease protection assay *(5)*. Once the hybridization is complete the single-stranded RNA of the sample and unhybridized probe are digested using a single-stranded ribonuclease (such as RNase A). It is believed that RNase Protection assays of this type are more sensitive than Northern analysis, yet there are serious limitations. These limitations relate to normalization between samples, which involves hybridization of a control target mRNA in the same tube as the experimental mRNA. The hybrids of both the control and experimental mRNA may be quantitated using gel electrophoresis, filter transfer, autoradiography, and then densitometry.

Low abundance mRNA species are still best examined using reverse transcription polymerase chain reaction (RT-PCR), as in theory at least only one molecule of mRNA needs to be present for the reaction to proceed. The amount of DNA amplified using RT-PCR depends on the amount of original starting mRNA *(6)*. The basic principles of mRNA quantitation discussed so far are also instrumental in RT-PCR. An internal control is required; it can be a gene that is constitutively expressed so that experimental and control samples are coamplified in the same tube or an external control can be used. Use of external controls is based on the principle of competitive RT-PCR, and it is this method that actually allows absolute quantitation of mRNA. Primer sets are designed to amplify the gene of interest and an artificial RNA specially designed to be a different size to the gene of interest. The availability of real-time PCR has further simplified the quantitation of competitive RT-PCR. However, this technique requires expensive equipment and the use of specific primers that have been fluorescently labeled.

The techniques discussed thus far function well if the tissue to be studied is of a homogeneous cell type; however, complications arise if the tissue is heterogeneous. Quantitation using the methods discussed in the preceding becomes impossible if the total amount of particular mRNA species do not change significantly, yet there are changes in cellular distribution of the mRNA or focal changes in expression levels. *In situ* hybridization is a technique that allows cellular localization of mRNA and the spatial distribution of cells expressing the gene of interest to be studied. In this way, quantitative data can be extracted from heterogeneous tissues where Northern blotting or RT-PCR would provide little information about changes in gene expression. Thus, in heterogeneous tissues or where there may be focal "hot spots" of expression of particular genes, *in situ* hybridization may be the best means of extracting quantitative information regarding levels of gene expression and their spatial distribution. An example of a very focal distribution of expression that can be quantitated by *in situ* hybridization is expression of the procollagen I gene during renal interstitial fibrosis, where expression in a large organ is localized to a small number of cells in the interstitium ***(7)***.

To obtain quantitative information regarding the amount of mRNA in a tissue section, computer-based image analysis systems can be used. This chapter examines the quantity of collagen I mRNA as represented by silver grains in skin wound tissue sections using Video Pro 32 © 1992–1995 (produced by Leading Edge Pty, South Australia, Australia) or Image Pro Plus (Media Cybernetics, Silver Spring, MD, USA). These software packages allow the extraction and measurement of image elements to provide a quantitative measurement of particular features. The extraction and storage of the image involves the conversion from a color image to its digital form. The video digitizer breaks the continuous color image into discrete pieces or pixels. Each pixel conveys color and brightness, and the digitizer creates values for each. It is these values from the color and brightness that the image analysis software uses to process and measure features of the image ***(8)***. Lastly, quantitation of *in situ* hybridization relies heavily on adequate controls being performed and on planning of experiments to provide sufficient numbers of samples to give reproducible and statistically valid results.

2. Materials

2.1. Tissue Preparation

1. Paraformaldehyde (Merck, Darmstadt, Germany).
2. Phosphate-buffered saline (PBS), pH 7.4: 0.14 *M* NaCl, 0.003 *M* KCl, 0.008 *M* Na_2HPO_4, 0.0015 *M* KH_2PO_4.
3. Ethanol, laboratory grade.
4. Chloroform (BDH, Poole, UK).
5. Paraplast (or similar) embedding wax (melting point 56°C).
6. Stainless steel embedding molds (Tissue Tek).
7. Glass vials for tissue processing.

2.2. Slide Preparation

1. Glass slides.
2. 3-Aminopropyltriethoxy-silane (APES) (Sigma, St. Louis, USA).
3. Acetone (BDH, Poole, UK).

2.3. Pretreatment of Paraffin-Embedded Tissue

All buffers are treated with 0.05% diethylpyrocarbonate (DEPC) (Sigma, St. Louis, MO, USA), with the exception of Tris-based buffers.

1. Histolene (Histolabs/Fronine, NSW, Australia) or xylene.
2. Ethanol.
3. PBS.
4. 1× Antigen retrieval solution (Citra) (BioGenex, CA, USA).
5. Pronase buffer (P buffer): 50 m*M* Tris-HCl, pH 7.5, 5 m*M* EDTA, pH 8.0.
6. Pronase E (protease from *Streptomyces griseus*) (Sigma, St. Louis, MO, USA).
7. 0.1 *M* Sodium phosphate buffer, pH 7.2.
8. 0.2 *M* HCl in water.
9. Double-distilled water, DEPC (Sigma, St Louis, MO. USA) treated.
10. 70% Ethanol.

2.4. Labeling of the Probe

1. cDNA in appropriate in vitro transcription vector providing polymerase sites for cRNA production (T7, T3, and/or SP6).
2. 5× Transcription buffer: 200 m*M* Tris-HCl, pH 7.5, 30 m*M* $MgCl_2$, 10 m*M* spermidine (Sigma, St. Louis, MO, USA), 50 m*M* NaCl.
3. 100 m*M* Dithiothreitol (DTT) (Boehringer Mannheim, Mannheim, Germany).
4. RNasin, ribonuclease inhibitor (Promega Corporation, WI, USA).
5. 10 m*M* ATP, 10 m*M* CTP, 10 m*M* GTP, 12 μ*M* UTP, (Promega Corporation, WI, USA).
6. RNA polymerases T7, T3, and SP6 (Promega Corporation, WI, USA).
7. Radionucleotide; 5'[α-^{33}P]UTP.
8. DNase I (Promega Corporation, WI, USA).
9. 20 mg/mL Stock transfer RNA (tRNA) (Boehringer Mannheim, Mannheim, Germany).
10. 7.5 *M* $NH_4C_2H_3O_2$.
11. 3 *M* $NaC_2H_3O_2 \cdot 3H_2O$, pH 5.2.
12. Ethanol.
13. Hydrolysis buffer: 80 m*M* $NaHCO_3$, 120 m*M* Na_2CO_3; 20 m*M* β-mercaptoethanol.
14. Stop buffer: 200 m*M* acetate buffer, pH 6.0, 1% glacial acetic acid, 10 m*M* DTT.
15. DEPC-treated double-distilled H_2O.
16. Dry heat block or water bath accurately set at 37°C.
17. Microcentrifuge.

2.5. Hybridization

1. 10× Salts: 3 *M* NaCl; 100 m*M* Na_2HPO_4, 100 m*M* Tris-HCl, pH 7.5, 50 m*M* EDTA, 0.2% bovine serum albumin, 0.2% Ficoll, 0.2% polyvinylpyrrolidone.

2. Formamide (BDH, Poole, UK).
3. Dextran sulfate (Amrad Pharmacia Biotech, Uppsala, Sweden).
4. tRNA.
5. DEPC-treated distilled H_2O.

2.6. Posthybridization Washes

1. 20x Standard saline citrate (SSC) solution, 3 *M* NaCl, 0.3 *M* sodium citrate.
1. Wash buffer: 2× SSC, 50% formamide.
2. RNase A (Sigma, St. Louis, MO, USA).
3. RNase buffer: 10 m*M* Tris-HCl, pH 7.5, 1 m*M* EDTA, 0.5 *M* NaCl.

2.7. Autoradiography and Emulsion Detection of Hybridization

1. X-Ray film cassette.
2. Film (XAR-5 or Hyperfilm).
3. Liquid nuclear research emulsion (gel form) (Ilford, Cheshire, UK).
4. Developer, Phenisol (Ilford), diluted 1:4 with distilled H_2O.
5. Hypam fixer (Ilford), diluted 1:4 with distilled H_2O.
6. Harris' hematoxylin stain.
7. Eosin stain.
8. Scott's tap water: 82 m*M* $MgSO_4{\cdot}7H_2O$, 42 m*M* $NaHCO_3$.
9. Mounting medium, nonaqueous.

2.8. Image Capture and Analysis

1. Microscope with image capture facility such as a charge coupled device camera (CCD) or digital camera.
2. Image analysis software such as VideoPro (Leading Edge Pty, South Australia, Australia) or Image Pro Plus (Media Cybernetics, Silver Spring, MD, USA).

3. Methods

3.1. Tissue Fixation, Processing, and Embedding

1. Place tissue biopsy in 4% paraformaldehyde–PBS, overnight at room temperature.
2. Wash the tissue in 7% sucrose/0.1 *M* sodium phosphate buffer overnight at 4°C.
3. Dehydrate tissue through graded alcohols (50%, 70%, 90%, 100%) and then two changes of 100% chloroform.
4. Place tissue in molten paraffin wax (approx 58°C) and leave tissue in wax for a minimum of 4 h.
5. Discard primary wax and replace with fresh wax and leave for 4 h.
6. Ensuring correct orientation of the tissue, embed the tissue in wax using stainless steel molds. Place the molds at –20°C for 1 h and then remove the wax block from the mold.

3.2. Coating Slides with APES

1. Place glass slides in racks and wash in an alkaline detergent overnight.
2. Rinse the slides thoroughly with running water and then allow them to dry.
3. Wrap the slides in aluminum foil and sterilize by baking at 180°C for 3 h (e.g., in a dry-heat sterilizer).
4. Place slides in racks and immerse in a 2% solution of APES in acetone for 20 s.
5. Rinse slides in acetone for 20 s and then in distilled H_2O, twice.
6. Dry the slides at 37°C overnight and store in an airtight container.

3.3. Tissue Sectioning

1. Fill a small container with distilled H_2O and prepare a water bath at 42°C.
2. Cut 5-μm sections of the paraffin embedded tissue on a microtome.
3. Place the sections into the H_2O and then with an uncoated glass slide transfer the section into the water bath. The section should flatten.
4. Mount the section with a coated slide and allow the section to dry overnight at 42°C.

3.4. Pretreatment of Tissue

1. Dewax the sections in histolene or xylene and rehydrate through graded alcohols and finally DEPC-treated distilled H_2O.
2. Microwave sections in 1× Citra solution, allow to cool slightly.
3. Rinse the sections in prewarmed (37°C) P buffer.
4. Digest tissue with Pronase E in P buffer (125 μg/mL) at 37°C for 10 min.
5. Rinse twice in 0.1 *M* sodium phosphate buffer.
6. Post fix the sections in 4% paraformaldehyde–PBS at room temperature for 10 min.
7. Rinse twice in 0.1 *M* sodium phosphate buffer.
8. Wash the sections in distilled H_2O and dehydrate in 70% ethanol twice.
9. Air-dry sections and store at room temperature in a closed container until required.

3.5. Labeling the Probe

Template concentration is important in the labeling procedure and for riboprobe synthesis 500–1000 ng of template is recommended.

1. For one transcription reaction the following final concentrations of reagents are required; 1× transcription buffer, 16 m*M* DTT, 20 U RNasin, 400 μ*M* ATP, 400 μ*M* CTP, 400 μ*M* GTP, 12 μ*M* UTP, template (500–1000 ng), 20 U of appropriate RNA polymerase, 50 μCi 5'[α-^{33}P]UTP, and distilled H_2O, to a final volume of 20 μL.
2. Incubate the reaction mixture at 37°C for 1 h in a dry heat block or water bath.
3. Digest the template DNA with 1 U of DNase I and incubate the reaction at 37°C for a further 15 min.

4. Add 40 μg of tRNA and adjust the reaction volume to 100 μL with DEPC-treated distilled H_2O.
5. Set aside 1 μL for scintillation counting.
6. Precipitate the riboprobe by adding 50 μL of 7.5 *M* $NH_4C_2H_3O_2$ and 300 μL of 100% ethanol and place at –70°C for 20 min.
7. Pellet the riboprobe by centrifugation at 10,000*g*, for 20 min at room temperature.
8. Remove the supernatant and wash the pellet with 70% ethanol.
9. Resuspend the riboprobe in 100 μL of DEPC-treated distilled H_2O and remove 1 μL for scintillation counting.
10. In the case of long probes, access to the target mRNA in the tissue may be limited. To improve penetration of the probe hydrolysis may be necessary. We have chosen a probe length of approx 0.15 kb. For hydrolysis, add 100 μL of hydrolysis buffer to the riboprobe and incubate at 65°C for the appropriate length of time (*see* **Note 1**).
11. Terminate the hydrolysis reaction by adding stop buffer and then precipitate the hydrolyzed probe by adding: 40 μL of 3*M* sodium acetate, 40 μg of tRNA, and 800 μL of 100% ethanol.
12. Precipitate as described in **steps 7** and **8**.
13. Resuspend in 100 μL of DEPC-treated distilled H_2O, and take 1 μL for scintillation counting.

3.6. Hybridization

1. Five hundred microliters of hybridization buffer consists of: 1× salts, 50% formamide, 10% dextran sulfate, 360 μg of tRNA. Five hundred microliters is sufficient for approx 10 sections.
2. Add labeled riboprobe to the hybridization buffer at a concentration of 20×10^6 per 500 μL of hybridization buffer.
3. Heat the probe–hybridization buffer mix to 85°C for 5 min before placing on the sections.
4. Coverslip the sections and place in a humidified airtight chamber.
5. Hybridize overnight at 60°C.

3.7. Posthybridization Washes

1. Heat wash buffer to 55°C and soak slides to remove coverslips.
2. Wash slides at 55°C for 30 min. Replace wash buffer and wash slides for a further 30 min.
3. Wash slides in three changes of RNase buffer and then incubate the sections with RNase A (150 μg/mL) in RNase buffer at 37°C for 1 h, with agitation (shaking water bath).
4. Wash the sections in 2× SSC for 45 min at 55°C and then dehydrate through graded alcohols and air-dry.

3.8. Autoradiography

1. Sections that have been hybridized with ^{33}P-labeled probes can be placed on X-ray film, to provide an idea of the success or otherwise of the hybridization reaction. This preliminary autoradiography can also serve as a guide for exposure times required in the liquid emulsion autoradiography step. However, for small pieces of tissue or where few cells are labeled in the tissue section, this step may be omitted.
2. In a darkroom under safelight illumination (Ilford safelight filter number 904 or Kodak safelight filter no. 2), weigh out 10 g of emulsion, add 6 mL of distilled H_2O, and incubate at 42°C for 2 h to allow the emulsion to melt.
3. Pour the liquefied emulsion into a dipping chamber (available from Amersham) and "dip" the experimental slides, ensuring all slides are coated evenly and that there are no air bubbles.
4. Remove excess emulsion by allowing slides to drain vertically on absorbent paper in the dark.
5. Place slides into a plastic slide rack and store in a lightproof box containing desiccant.
6. Expose in the lightproof container for 10–20 d, depending on the strength of the hybridization signal.

3.9. Signal Development

1. In a darkroom under safelight illumination, place slides in the diluted developer for 2 min with mild agitation.
2. Stop development by immersion in 0.5% acetic acid for 30 s.
3. Immerse the slides for 2 min in rapid fixer.
4. Rinse slides in running tap water for 5 min.
5. Stain slides with Harris' hematoxylin, rinse in tap water, and place in Scott's tap water for 30 s or until hematoxylin appears blue; rinse in water and then stain with eosin.
6. Dehydrate sections through graded alcohols, rinse in two changes of histolene and mount using a nonaqueous mountant.

3.10. Quantitation of In Situ Hybridization on Sections

1. Once the image analysis system is set up and a live image is viewed on the monitor, the image can be captured by the analysis program. This function can be selected from the drop-down menu in most image analysis programs. The monitor now displays a frozen image of your specimen, which is the live image in a digitized form. Alternatively, in the case of the Video Pro system images stored as bitmaps (.bmp) may be imported into the program, or in Image Pro Plus .jpg or .tif files may be used.
2. Choose the *select* function from the color menu; this will permit a color to be selected from the image (in this case the black of silver grains, *see* **Note 2**) and for this color to be distinguished from other features in the section. In most image

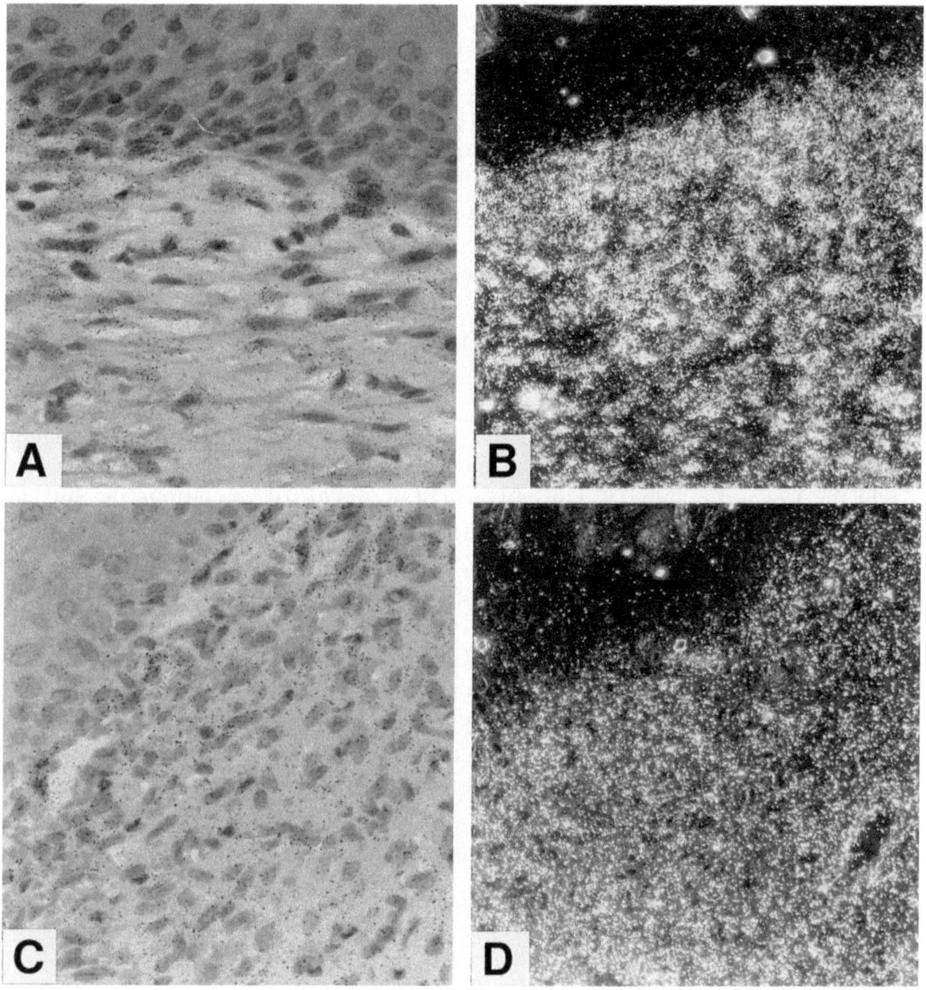

Fig. 1. (a) *In situ* hybridization for $\alpha 1$ (I) procollagen in a section of control mouse wound tissue 10 d postwounding. Fibroblasts in the granulation tissue strongly express the gene during tissue repair. **(b)** Darkfield micrograph of the same section shown in **(a)**. **(c)** In wound tissue from a diabetic mouse, $\alpha 1$ (I) procollagen mRNA expression is markedly lower. (d) Darkfield micrograph of the same tissue section shown in (c). Grain counts of the two tissue taken across ten $\times$ 40 magnification fields showed expression was approximately threefold higher in the control tissue (control, 2971 ± 244 grains/field compared to diabetic tissue, 1059 ± 323 grains/field).

analysis packages there is a command that allows the area of interest to be delineated with the cursor and for this color to then be quantitated by the *count* or *measure* command.

3. Adjustment of the color and detection limits can be modified at this point to maximize the distinction of silver grains from background colors in the sample. This produces an overlay on the section with the areas of interest highlighted. Once a set of color threshold parameters are chosen they then need to be consistently applied to all the sections to be studied.
4. Difficulty arises when the color discriminated is similar to another feature that the operator is not measuring (such as very darkly stained nuclei). If these features differ in size compared to the features of interest then the *sizer* function from the *measure* menu allows for exclusion of features above or below a particular size (in pixels) (*see* **Note 3**).
5. It is at this point that the automatic sequential measurement of the individual features in the image according to size can be achieved.
6. In the Image Pro Plus system, *count* will automatically quantitate the selected features at this point, as will the *measure* command in the Image Pro system.
7. We generally use a fixed number of 20× or 40× magnification fields depending on the size of the sample being viewed. In this way a statistically reliable measure of grain numbers per field can be obtained and comparisons made between treatments. An example of sections labeled with a α1 (I) procollagen probe is given in **Fig. 1**. Ten 40× magnification fields were captured and the grains quantitated for each condition (diabetic vs nondiabetic control). Grain count ± standard deviation is given. In general, sections would be counted from at least five animals in each group. To avoid differences across hybridization batches, comparisons are made only between groups of sections, which were hybridized and treated in the same sample run (*see* **Note 4**).

4. Notes

1. Hydrolysis formula:

$$\text{hydrolysis time (min)} = \frac{\text{length of probe (kb)} - \text{length of desired end product (kb)}}{[0.11 \times \text{length of probe (kb)} \times \text{length of desired end product (kb)}]}$$

For example, starting with a probe that is 1.5 kb and requiring an end product of 0.15 kb, the hydrolysis time is:

$$t = \frac{1.5 - 0.15}{0.11 \times 1.5 \times 0.15} = 54.5 \text{ min}$$

2. Another option is to use darkfield microscopy where silver grains will appear as white spots on a dark background. In the Image Pro Plus software package, silver grains can be automatically selected in darkfield by *automatic bright object selection* and then quantitated using the *count* command.

3. An upper figure can be given in most programs above that features will be excluded from the count. This allows the user to exclude other artifacts, such as the occasional large clump of grains or nuclei if these are highly stained and are recognized as the same color as silver grains. In some cases grains representing specific labeling may be partially clumped together, and are therefore counted as a single object rather than single grains. When this occurs, the *erode* command in image analysis packages can be used to separate grains.
4. To achieve statistically valid results, sufficient numbers of sections from each treatment or each experimental group need to be hybridized in the same batch. Because probe labeling efficiency, incubation conditions, developing temperatures, etc. may all lead to slight variations in results, it is virtually impossible to sufficiently standardize all steps of the hybridization, and detection procedures, to compare sections between *in situ* hybridization runs that were carried out at different times. Grain counts should also be adjusted for nonspecific background labeling of the section that occurs with isotopic *in situ* hybridization. In tissues where there are focal areas of labeling, adjacent unlabeled areas may be used for correcting counts. Adjacent sections labeled with sense probes are not as useful for this purpose, as labeling efficiency and subsequent background labeling may not be the same as for the antisense probe. Where precise measures of labeling efficiency and specific activity can be achieved, sense probes may be used for calculation of background or nonspecific labeling.

References

1. Alwine, J. C., Kemp, D. J., and Stark, G. R. (1977) Method for detection of specific RNA's in agarose gels by transfer to diazobenzyloxymethyl-paper and hybridization with DNA probes. *Proc. Natl. Acad. Sci. USA* **74,** 5350–5354.
2. Thomas, P. S. (1980) Hybridization of denatured RNA and small DNA fragments transferred to nitrocellulose. *Proc. Natl. Acad. Sci. USA* **77,** 5201–5205.
3. Reue, K. (1998) mRNA quantitation techniques: considerations for experimental design and applications. *J. Nutr.* **128,** 2038–2044.
4. White, B. A. and Bankcroft, F. C. (1982) Cytoplasmic dot hybridization. *J. Biol. Chem.* **257,** 8569–8572.
5. Bark, A. J. and Sharp, P. A. (1977) Sizing and mapping of early adenovirus mRNA's by gel electrophoresis of S1 endonuclease digested hybrids. *Cell* **12,** 721–732.
6. Foley, K. P., Leonard, M. W., and Engel, J. D. (1993) Quantitation of RNA using the polymerase chain reaction. *Trends Genet.* **9,** 380–385.
7. Hewitson, T. D., Darby, I. A., Bisucci, T., Jones, C. L., and Becker, G. J. (1998) Evolution of tubulointerstitial fibrosis in experimental renal infection and scarring. *J. Am. Soc. Nephrol.* **9,** 632–642.
8. Leading Edge Pty, Ltd. (1992–1995) Video Pro 32 Reference manual.

14

Stable Isotope Probing for Detection of Methanotrophs After Enrichment with $^{13}CH_4$

Stefan Radajewski and J. Colin Murrell

1. Introduction

Methanotrophic bacteria are a physiologically specialized group of microorganisms that can use methane as a sole source of carbon and energy. They can be isolated from a wide range of environments including soil, sediment, peat bogs, and hot springs *(1–4)*, although they are often relatively difficult to maintain in laboratory culture. Characterization of the extant methanotrophs has identified several common features of their taxonomy and biochemical pathway for methane oxidation (reviewed in *[5]*) that are briefly summarized here.

The nine recognized methanotrophic genera are divided into two groups based on 16S rRNA phylogeny. The genera *Methylococcus*, *Methylomonas*, *Methylobacter*, *Methylomicrobium*, *Methylocaldum*, and *Methylosphaera*, which are known as type I methanotrophs, form a closely related clade within the γ-subclass of the *Proteobacteria*. The type II methanotrophic genera include *Methylosinus, Methylocystis*, and *Methylocella*, which form a closely related clade within the α-subclass of the *Proteobacteria*.

All methanotrophs use the enzyme methane monooxygenase (MMO) to catalyze the oxidation of methane to methanol. There are two distinct forms of MMO: a membrane-bound, particulate form (pMMO) and a cytoplasmic, soluble form (sMMO) (reviewed in *[6]*). One or both of these forms of MMO are found in all known methanotrophs, with pMMO reported in all genera except *Methylocella (7)*. The second enzyme in the pathway of methane oxidation is methanol dehydrogenase (MDH), which is found in most Gram-negative methylotrophs, including all methanotrophs *(8)*. The genes encoding key

From: *Methods in Molecular Biology, vol. 179: Gene Probes: Principles and Protocols*
Edited by: M. Aquino de Muro and R. Rapley © Humana Press Inc., Totowa, NJ

enzymes involved in a specific function, such as methane metabolism, are known as "functional genes." Characterization of the phylogenetic and functional genes of extant methanotrophs has resulted in the development of a range of molecular biological techniques, centered on gene probes, for studying methanotroph ecology (reviewed in *(9)*).

The technique of stable isotope probing (SIP) *(10)* described here identifies the active members of a functionally specialized group of microorganisms under conditions that approach those occurring *in situ*. Therefore, SIP is ideally suited to studying methanotroph ecology because it does not require cultivation of the microorganisms, but exploits the wide range of gene probes that are available to characterize the functionally active population of methanotrophs. Three steps are involved in SIP:

1. Labeling of methanotrophic cells with ^{13}C from isotopically enriched $^{13}CH_4$
2. Separation of ^{13}C-labeled ("heavy") DNA from ^{12}C-labeled ("light") DNA
3. Analysis of [^{13}C]DNA using PCR and gene probes to determine the identity of the active methanotrophs.

A unique feature of SIP is the isolation of DNA from the functionally active microorganisms after growth on a compound enriched with a stable isotope such as ^{13}C. The basis of this technique lies in the natural abundance of ^{13}C being only 1.1%. Consequently, the addition of an isotopically enriched compound such as $^{13}CH_4$ (99% atom enriched) to an environmental sample will result in actively dividing methanotrophs becoming labeled with ^{13}C. The DNA of the active methanotrophs will therefore become ^{13}C-labeled ("heavier"), enabling it to be separated from [^{12}C]DNA of nonmethanotrophs by centrifugation in a CsCl–ethidium bromide density gradient.

2. Materials

1. $^{13}CH_4$ (99% ^{13}C; Linde Gas, UK or CK Gas Products, UK).
2. Crimp topped serum vials (125 mL), chlorobutyl rubber seals, and aluminum crimps (Adelphi, UK).
3. Gas chromatograph equipped with a flame ionization detector. Using a 1 m length × 4 mm inner diameter glass column containing Poropak Q (Sigma), a column temperature of 150°C, and a carrier gas (N_2) flow rate of 25 mL/min^{-1}, the retention time of CH_4 is 0.33 min.
4. Bead beater (B. Braun, Melsungen AG, Germany) and glass beads (0.10–0.11 mm diameter; B. Braun Biotech International GmbH, Germany).
5. FAST-PREP DNA extraction kit (BIO 101, USA).
6. 0.1 *M* Sodium phosphate buffer, pH 7.0, sterilized and stored at room temperature.
7. Binding matrix: Glassmilk (BIO 101, USA) diluted in a 1:1 ratio in 6 *M* guanidine HCl.

8. Wash buffer: 70% (v/v) ethanol containing 0.1 *M* sodium acetate.
9. CsCl solution: 1 g of CsCl dissolved in 1 mL of TE.
10. Ethidium bromide: 10 mg/mL in water.
11. Ultracentrifuge equipped with a VTi 65 or VTi65.2 rotor (Beckman, USA).
12. Polyallomer Quick-Seal centrifuge tubes (13 mm × 51 mm) (Beckman, USA).
13. Ammonium acetate: 10 *M*, filter sterilized (0.2 μm) and stored at room temperature.
14. Dialysis membrane tubing is prepared by boiling lengths (~10–20 cm) for 10 min in a large volume of 2% (w/v) $NaHCO_3$ and 1 m*M* EDTA, pH 8.0. Tubing is rinsed thoroughly in distilled water and boiled for 10 min in 1 m*M* EDTA (pH 8.0). Store tubing at 4°C and wash thoroughly with distilled water before use.
15. TE: 10 m*M* Tris-HCl, pH 7.6; 1 m*M* disodium EDTA.
16. *Taq* DNA polymerase, dNTPs, PCR primers (*see* **Table 1**), and reaction buffers (supplied by Life Technologies, UK and others).
17. TOPO TA cloning kit (Invitrogen, USA).

3. Methods

3.1. Labeling of Methanotrophs with ^{13}C

Incorporation of sufficient ^{13}C into the DNA of the active methanotrophs is critical for the success of SIP (*see* **Note 1**). Therefore, it is worth determining the potential of the sample to oxidize methane ($^{12}CH_4$) before addition of $^{13}CH_4$. Furthermore, as preferential use of $^{13}CH_4$ instead of other substrates is required for effective separation of DNA, this technique is best suited to higher methane concentrations (e.g., >1% v/v). A general method that has been broadly successful for labeling methanotroph DNA is described here, although some optimization may be required for certain environments (*see* **Note 2**). A separate control sample exposed to $^{12}CH_4$ is processed identically to the $^{13}CH_4$ sample as it enables a comparative analysis of the microbial diversity (sequence diversity) present in the sample following enrichment with methane.

1. The sample being investigated (soil, 10 g; slurries or aquatic/marine, 10 mL) is transferred to a serum vial, which is crimp sealed, injected with 10 mL of $^{13}CH_4$ (*see* **Note 3**) and incubated at the desired temperature.
2. The methane concentration is measured at regular intervals by gas chromatography. After the methane concentration has decreased to <5% of the starting concentration the vials are opened and flushed with air (~500 mL) to keep the headspace gas aerobic. Flushing with air also removes $^{13}CO_2$ produced during the oxidation of $^{13}CH_4$ (*see* **Note 4**). The vial is resealed and spiked with a further 10 mL of $^{13}CH_4$.
3. Successful labeling of DNA has been achieved after only 10 mL of $^{13}CH_4$ has been consumed. However, if **step 2** is repeated until 30–50 mL of $^{13}CH_4$ (~1–1.5 mmol ^{13}C) has been consumed, higher yields of [^{13}C]DNA are obtained.

3.2. Extraction of [^{13}C]DNA

Many laboratories will have preferred methods for extraction of genomic DNA from environmental samples. Nevertheless it is worth stressing that the method used must lyse as diverse a range of microorganisms as possible. The bead-beating method we describe is suitable for DNA extraction from soil samples and is adapted from Yeates and Gillings ***(18)***. Samples with a low particulate content (e.g., aquatic/marine) can be concentrated by centrifugation and lysed with the FAST PREP DNA extraction kit as described elsewhere ***(18)***. The [^{12}C]DNA is collected in an identical manner from the parallel $^{12}CH_4$ control sample.

1. Soil (3 g) is added to a bead beating tube with 1 g of sterile glass beads, 3 mL of sodium phosphate buffer, and 1 mL of sodium dodecyl sulfate (SDS) (10% w/v). The jacket is cooled with CO_2 vapor during a 5-min homogenization. The slurry is kept on ice while DNA is extracted in the same manner from the remainder of the sample.
2. The pooled slurries are centrifuged at 15,000 *g*, 5 min and the supernatant is transferred to a fresh centrifuge tube. To precipitate proteins, 0.2× volume of potassium acetate (7.5 *M*) is added, mixed gently, and centrifuged at 15,000 *g* for 5 min.
3. For soil samples it is necessary to purify the DNA further because humic material often hinders the observation of DNA in the ultracentrifuge tubes. The supernatant is added to a fresh centrifuge tube to which an equal volume of binding matrix is added. The sample is rotated gently but continually for 5 min and centrifuged at 15,000 *g* for 5 min. The binding matrix is gently resuspended in 1× volume of wash buffer, centrifuged for 1 min at 15,000 *g*, and the wash buffer discarded. The wash step is repeated 3× in total and 1 mL of TE is added to the pellet after the final centrifugation step. After standing for 1 min, the sample is centrifuged at 15,000 *g* for 5 min and the supernatant, containing the DNA, is collected. The elution step is repeated once to maximize DNA recovery.
4. The DNA solution is prepared for density gradient centrifugation by the addition of CsCl in the ratio 1 g CsCl/1 mL of DNA solution, and shaken gently until dissolved. Ethidium bromide (100 µL) is added to each ultracentrifuge tube, which is filled with CsCl solution and sealed. Tubes are centrifuged at 265,000 *g* (VTi 65; 55,000 rpm) for 12-16 h at 20°C.

Typical gradients of DNA from a methanotroph grown on $^{12}CH_4$ or $^{13}CH_4$ are shown in **Fig. 1a**. The ^{13}C-labeled DNA fraction appears as a distinct dense ("heavy") band below the [^{12}C]DNA. When working with environmental samples, a range of DNA fractions with intermediate levels of ^{13}C incorporation can be observed as a smear in a CsCl–EtBr density gradient (**Fig. 1b**). In such cases, the most labeled fraction can be collected using a modification of the centrifugation conditions (*see* **Note 5**).

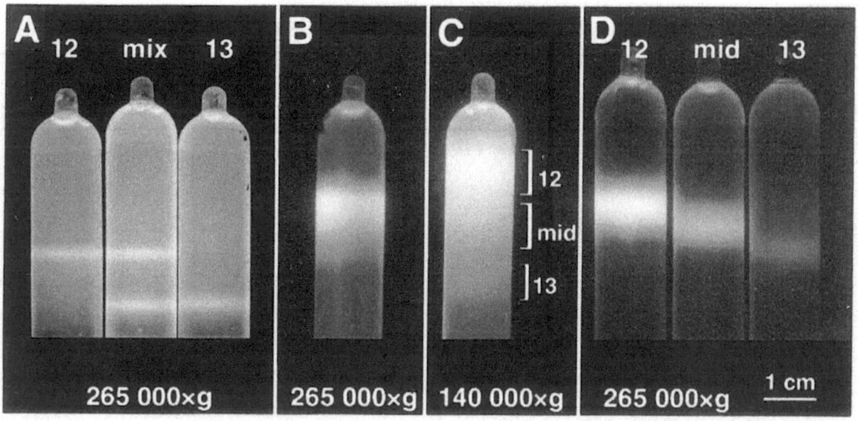

Fig. 1. Equilibrium centrifugation of isotopically labeled DNA in CsCl gradients. **(a)** Pure fractions and a mixture of the DNA extracted from a *Methylosinus trichosporium* culture utilizing either [^{12}C] or [^{13}C]methane as the sole carbon source. **(b)** DNA extracted from an environmental sample that had utilized $^{13}CH_4$, with a range of ^{13}C incorporation into DNA appearing as a smear. **(c)** Centrifugation at 140,000*g* for 60 h increases the separation of ^{12}C—and [^{13}C]DNA fractions, **(d)** facilitating the collection of discrete fractions with different levels of ^{13}C incorporation.

5. Collection of the ^{13}C-labeled fraction involves two ultracentrifugation steps. The primary step involves collection of the ^{13}C-labeled DNA fraction with a syringe (1 mL) and needle (19 g). The centrifuge tube is pierced with the needle 2 mm below the most dense DNA fraction, and approx 0.5 mL of CsCl solution containing the DNA is withdrawn gently and transferred to a microcentrifuge tube.
6. The ^{13}C-labeled DNA fraction is transferred to a new CsCl–ethidium bromide gradient and recentrifuged to purify the ^{13}C-labeled fraction from the small proportion of [^{12}C]DNA that can be inadvertently collected during the primary extraction. The second collection step is carried out identically to the first, taking care to collect the minimum volume possible (<0.5 mL).
7. Ethidium bromide is extracted from DNA by the addition of an equal volume of 1-butanol saturated with TE, followed by gentle mixing and brief centrifugation at 13,000*g*. The organic layer is discarded and the extraction repeated until the pink colour has been removed from the aqueous phase. The DNA is transferred to dialysis tubing that is sealed with dialysis clips. Dialysis is carried out for >2 h at 4°C against a large volume of TE (1–2 L) and is repeated 2–3×.
8. Dialyzed DNA is transferred to a microcentrifuge tube and precipitated overnight at –20°C by addition of 1/3× volume of ammonium acetate and 2× volume of ethanol. The DNA is pelleted by centrifugation at 13,000*g* for 20 min, washed with 2× volume of 70% ethanol, centrifuged at 13,000*g* for 10 min, resuspended in 50–100 µL of TE, and stored at –70°C.

Table 1
Methanotroph Functional and Phylogenetic Group-Specific Probes

Primer	Sequence (5'–3')[a]	Temp. (°C)[b]	Target gene or genus[c]	Reference
27f	AGAGTTTGATCMTGGCTCAG	60	16S rRNA	*11*
1492r	TACGGYTACCTTGTTACGACTT	60	16S rRNA	*11*
Mb1007r	CACTCTACGATCTCTACAG	58	*Methylobacter*	*12*
Mc1005r	CCGCATCTCTGCAGGAT	54	*Methylococcus*	*12*
Mm835	GCTCCACYACTAAGTTC	55	*Methylomonas*	*13*
Type2b	CATACCGGRCATGTCAAAAGC	55	*Methylosinus* *Methylocystis*	*13*
A189	GGNGACTGGGACTTCTGG	56	*pmoA*	*14*
A682	GAASGCNGAGAAGAASGC	56	*pmoA*	*14*
Mb661	CCGGMGCAACGTCYTTACC	55	*pmoA*	*13*
mmoX f882	GGCTCCAAGTTCAAGGTCGAGC	55	*mmoX*	*15*
mmoX r1403	TGGCACTCGTAGCGCTCCGGCTCG	55	*mmoX*	*15*
mmoX1	CGGTCCGCTGTGGAAGGGCATGAAGCGCGT	60[d]	*mmoX*	*16*
mmoX r901	TGGGTSAARACSTGGAACCGCTGGGT	60	*mmoX*	*17*
mxa f1003	GCGGCACCAACTGGGGCTGGT	55	*mxaF*	*12*
mxa r1561	GGGCAGCATGAAGGGCTCCC	55	*mxaF*	*12*

[a] N = A, C, G or T; M = A or C; R = A or G; S = C or G; Y = C or T.

[b] Annealing temperature for PCR.

[c] *pmoA*, *mmoX*, and *mxaF* are the genes coding for the active site subunits of pMMO, sMMO, and MDH, respectively.

[d] Annealing temperature altered from 55°C to 60°C by Shigematsu et al. *(17)*.

3.3. Molecular Analysis of [^{13}C]DNA for Identification of Methanotrophs

Molecular analysis of [^{13}C]DNA uses the established technology of the polymerase chain reaction (PCR). Sequences retrieved from the labeled DNA can be compared with the extensive database (GenBank; http://www.ncbi.nlm.nih.gov) of methanotroph sequences (both 16S rRNA and functional genes) and phylogenies reconstructed. As ^{13}C-labeled DNA must have been derived from microorganisms that were functionally active *in situ*, novel groups of methanotrophs can be detected by 16S rDNA analysis. Strategies can then be formulated to isolate specific genera of active methanotrophs using information available for the closest cultivated representatives.

Methanotroph-specific PCR primers are listed in **Table 1**. Specific amplification protocols are detailed in each reference. A 10-fold dilution series of DNA is used as a template for PCR amplifications, with an optimal amount of 5–50 ng per reaction. Typical amplification protocols involve an initial denaturation step at 94°C for 5 min, followed by addition of the *Taq* DNA polymerase, and 30 cycles of denaturation (94°C) for 1 min, annealing (*see* **Table 1**) for 1 min and extension (72°C) for 1 min. A final extension cycle for 10 min is used prior to cloning. All commercial kits and reagents are used according to manufacturer's recommendations.

1. PCR amplification of 16S rRNA genes initially involves universal eubacterial (27f and 1492r) primers to assess total bacterial diversity within the ^{13}C-labeled DNA. Methanotroph group-specific primers (Mb1007r, Mm835, Mc1005r, type 2b) can be used in conjunction with primer 27f to target known methanotrophic groups, which may be minor components of the active population.
2. PCR amplification of functional genes involves MMO—and MDH-specific primer sets; for *pmoA* (A189/A682 or A189/Mb661), for *mmoX* (mmoX f882/mmoX r1403 or mmoX1/mmoX r901) and for *mxaF* (mxa f1003/mxa r1561).
3. Clone libraries are constructed from amplification products using the TOPO TA cloning kit or equivalent (*see* **Note 6**). Libraries should include as many clones as is feasible to screen, especially for 16S rRNA genes (>50 clones).
4. Clone inserts are screened by digestion with restriction endonucleases. The enzyme *Eco*RI releases the insert from the cloning vector and can be used in combination with a range of other enzymes including *Rsa*I, *Sau*3AI, *Hin*cII, *Pvu*II, and others. Digested plasmids are screened on 2% (w/v) agarose gels and clones are grouped according to the restriction pattern.
5. Representative clones are sequenced using commercially available automatic sequencing facilities and phylogenetic analysis is undertaken using appropriate software (e.g., PHYLIP *[19]*, ARB *[20]*, and others).

4. Notes

1. In some soil samples we have been unable to observe a "heavy" ^{13}C-labeled DNA fraction, despite rapid methane oxidation by that soil. This may be due to cooxidation of CH_4 by other microorganisms *(21)*, thereby preventing assimilation of ^{13}C into methanotroph DNA with CH_4 as the substrate. Alternatively, the methane oxidizers may also be using other carbon substrates, thereby diluting the proportion of ^{13}C incorporated into DNA. Methanogenesis (production of $^{12}CH_4$) would also dilute the $^{13}CH_4$, although this is less likely in aerobic systems.
2. In the above example, a [^{13}C]DNA fraction was observed following the addition of dilute NMS (nitrate mineral salts medium in Whittenbury et al. *[1]*) to produce a soil slurry (4 g of soil, 10 mL of $1/10^{th}\times$ NMS) that was shaken at 20°C. Although it is not entirely representative of natural *in situ* conditions, a subgroup of methanotrophs was identified as being active in this complex environment.
3. Because of the relatively high cost of $^{13}CH_4$, the following methods can be used to recover the residual methane from their containers. Methane collected from a break-seal glass flask is displaced by the addition of the equivalent volume of water. When methane supplied in steel cylinders is at atmospheric pressure, the cylinder can be heated in a water bath to expand the remaining gas.

4. All ^{13}C-labeled intermediates/products of metabolism, including $^{13}CH_3OH$, $H^{13}CHO$, $H^{13}COOH$, and $^{13}CO_2$ may potentially act as substrates for other microorganisms if excreted from methanotroph cells. The final product of $^{13}CH_4$ oxidation ($^{13}CO_2$) can be removed by flushing serum vials with air after methane consumption or by inclusion of a CO_2 trap (e.g., NaOH solution).
5. Centrifugation at slower speed (140,000*g* [VTi 65, 40,000 rpm], 60 h, 20°C) will narrow the density gradient, thereby increasing the separation of DNA fractions at the expense of less sharp band formation (**Fig. 1c**). This facilitates easier collection of discrete fractions from the "stretched" gradient (**Fig. 1d**).
6. With some MMO primer sets, nonspecific PCR products can be observed. To improve the efficiency of cloning the target gene, PCR products should be separated by electrophoresis through a 1.2% (w/v) agarose gel. DNA is visualized using an ultraviolet transilluminator and the correct sized product is excised, purified using a gel purification kit (e.g., Gene Clean, BIO 101, USA), and cloned.

References

1. Whittenbury, R., Phillips, K. C., and Wilkinson, J. F. (1970) Enrichment, isolation and some properties of methane-utilizing bacteria. *J. Gen. Microbiol.* **61,** 205–218.
2. Bowman, J. P., Sly, L. I., Nicholas, P. D., and Hayward, A. C. (1993) Revised taxonomy of the methanotrophs: description of *Methylobacter* gen. nov., emendation of *Methylococcus*, validation of *Methylosinus* species and a proposal that the family *Methylococcaceae* includes only group I methanotrophs. *Int. J. Syst. Bacteriol.* **43,** 735–753.
3. Bodrossy, L., Murrell, J. C., Dalton, H., Kalman, M., Puskas, L. G., and Kovacs, K. (1995) Heat-tolerant methanotrophic bacteria from the hot water effluent of a natural gas field. *Appl. Environ. Microbiol.* **61,** 3549–3555.
4. Dedysh, S. N., Panikov, N. S., Liesack, W., Großkopf, R., Zhou, J., and Tiedje, J. M. (1998) Isolation of acidophilic methane-oxidizing bacteria from northern peat wetlands. *Science* **282,** 281–284.
5. Hanson, R. S. and Hanson, T. E. (1996) Methanotrophic bacteria. *Microbiol. Rev.* **60,** 439–471.
6. Murrell, J. C., McDonald, I. R., and Gilbert, B. (2000) Regulation of expression of methane monooxygenases by copper ions. *Trends Microbiol.* **8,** 221–225.
7. Dedysh, S. N., Liesack, W., Khmelenina, V. N., Suzina, N. E., Trotsenko, Y. A., Semrau, J.D., et al. (2000) *Methylocella palustris* gen. nov., sp. nov., a new methane-oxidizing acidophilic bacterium from peat bogs, representing a novel subtype of serine-pathway methanotrophs. *Int. J. Syst. Evol. Microbiol.* **50,** 955–969.
8. Anthony, C. (1986) Bacterial oxidation of methane and methanol. *Adv. Microb. Physiol.* **27,** 113–210.
9. Murrell, J. C., McDonald, I. R., and Bourne, D. G. (1998) Molecular methods for the study of methanotroph ecology. *FEMS Microbiol. Ecol.* **27,** 103–114.
10. Radajewski, S., Ineson, P., Parekh, N. R., and Murrell, J. C. (2000) Stable-isotope probing as a tool in microbial ecology. *Nature* **403,** 646–649.

11. Lane, D. J. (1991) 16S/23S rRNA sequencing. In *Nucleic Acid Techniques in Bacterial Systematics* (Stackebrandt, E. and Goodfellow, M., eds.), Academic Press, Chichester, pp. 115-175.
12. Holmes, A. J., Owens, N. J. P., and Murrell, J. C. (1995) Detection of novel marine methanotrophs using phylogenetic and functional gene probes after methane enrichment. *Microbiology* **141,** 1947–1955.
13. Costello, A. M. and Lidstrom, M. E. (1999) Molecular characterization of functional and phylogenetic genes from natural populations of methanotrophs in lake sediments. *Appl. Environ. Microbiol.* **65,** 5066–5074.
14. Holmes, A. J., Costello, A., Lidstrom, M. E., and Murrell, J. C. (1995) Evidence that particulate methane monooxygenase and ammonia monooxygenase may be evolutionarily related. *FEMS Microbiol. Lett.* **132,** 203–308.
15. McDonald, I. R., Kenna, E. M., and Murrell, J. C. (1995) Detection of methanotrophic bacteria in environmental samples with the PCR. *Appl. Environ. Microbiol.* **61,** 116–121.
16. Miguez, C. B., Bourque, D., Sealy, J. A., Greer, C. W., and Groleau, D. (1997) Detection and isolation of methanotrophic bacteria possessing soluble methane monooxygenase (sMMO) genes using the polymerase chain reaction (PCR). *Microb. Ecol.* **33,** 21–31.
17. Shigematsu, T., Hanada, S., Eguchi, M., Kamagata, Y., Kanagawa, T., and Kurane, R. (1999) Soluble methane monooxygenase gene clusters from trichloroethylene-degrading *Methylomonas* sp. strains and detection of methanotrophs during in situ bioremediation. *Appl. Environ. Microbiol.* **65,** 5198–5206.
18. Yeates, C. and Gillings, M. R. (1998) Rapid purification of DNA from soil for molecular biodiversity analysis. *Lett. Appl. Microbiol.* **27,** 49–53.
19. Felsenstein, J. (1993). PHYLIP—Phylogeny Inference Package (Version 3.5c).
20. Strunk, O., Westram, R., Gross, O., Reichel, B., May, M., Hermann, S., et al. (1998). ARB: a software environment for sequence data. [online]. http://www.mikro.biologie.tu-muenchen.de.
21. Hyman, M. R., Murton, I. B., and Arp, D. J. (1988) Interaction of ammonia monooxygenase from *Nitrosomonas europaea* with alkanes, alkenes, and alkynes. *Appl. Environ. Microbiol.* **54,** 3187-3190.

15

Detection of *Enterococci* in Freshwater and Seawater (16S and 23S rRNA *Enterococcus* Oligonucleotide Probes)

Gillian D. Lewis, Sally A. Anderson, and Susan J. Turner

1. Introduction

Enteric bacteria, such as members of the enterococci and fecal coliform groups, are widely used as sanitary indicators of water quality in marine and freshwater systems. However, recent studies investigating the survival characteristics of these enteric bacteria in the laboratory ***(1–3)***, and *in situ* ***(4,5)*** have indicated that traditional enumerative techniques, based on selective culture, do not detect all viable bacteria present. It is now widely acknowledged that microorganisms become injured as a result of exposure to stressful environments ***(6)*** and that this injury may impede growth under the stringent conditions imposed by selective culture. The term "viable but nonculturable" (VNC) has been used to describe those organisms that, on exposure to an adverse environment ***(7–9)***, can no longer be cultured but have maintained some metabolic capability, as evidence by vital staining ***(10–12)***, and in some cases have retained their virulence ***(7,13,14)***. The VNC state has been demonstrated for a number of bacterial species including *E. coli* ***(9,15)***, and enterococci ***(16)***, and occurs in response to a range of environmental stresses such as sunlight exposure, salinity, nutrient limitation, and pH and temperature fluctuations.

The entry of these bacteria into a VNC state has significant implications for water quality monitoring ***(4,5,17–19)*** because traditional enumeration methods may significantly underestimate the actual number of viable organisms present in a sample. Culture-based enumeration methods may therefore yield results that vary depending on the indicator organism's exposure to environmental stress. While reducing the culturable count of indicator bacteria, these

From: *Methods in Molecular Biology, vol. 179: Gene Probes: Principles and Protocols*
Edited by: M. Aquino de Muro and R. Rapley © Humana Press Inc., Totowa, NJ

sublethal stressors may not affect the occurrence or virulence of microbial pathogens. Thus stress-induced loss of culturability is a fundamental problem in water quality assessment.

It is well known that injured or nonculturable cells do have the ability to repair given the proper environment, for example, by enrichment or recovery on nonselective media. Thus one strategy for overcoming the VNC problem is to include a preliminary resuscitation step on nonselective media, followed by selective growth or identification of target species by direct methods such as gene probes ***(20)***. Specific gene probes have been developed for bacterial indicators including Enterococci ***(21-32)*** and *E. coli* ***(24)*** and such an approach has been used successfully for detecting these organisms in marine and freshwaters ***(25)***. In developing the *uid*A gene probe Green et al. (1991)***(24)*** noted that the detection of seawater stressed *E. coli* after resuscitation on a nonselective culturing media was 3 and 11 times more effective than detection using traditional culturing methodologies such as the mTEC membrane filtration method ***(26)***. From this observation and from other research ***(25,27)*** it would appear that a higher proportion of injured or nonculturable cells in marine and freshwater samples can be detected using gene probe methods.

The method presented in this chapter describes the procedure for the detection of *Enterococcus* species from aquatic environments. The Enterococci oligonucleotide gene probes are targeted to 16S or 23S rRNA gene sequences. The advantage of using rRNA genes as hybridization targets for synthetic oligonucleotides is the flexibility of probe specificity ***(28)***. By targeting regions of greater or lesser conservation, probes can be designed to be specific at a kingdom-, genus-, and species-specific level ***(20,28,29)***. The specific oligonucleotide probes described in this protocol are for the detection of *Ent. faecalis*, *Ent. faecium*, which are members of the *Enterococcus* group. A universal *Eubacterial* probe was also used as a positive control for hybridization. Probe sequences for *Ent. faecalis* and *Ent. faecium* are based on those initially described by Betzl et al. (1990) ***(21)*** with later modifications by Beimfohr et al. (1993) ***(30)***, and are targeted to variable regions of the 23S rRNA. The *Enterococcus*-specific probe ***(23)*** and the Eubacterial probe ***(28)*** target regions of the 16S rRNA. These oligonucleotide probes will also hybridize to the genes encoding these sequences at the DNA level. In a more recent publication Frahm et al. (1998) ***(22)*** described a number of additional 23S rDNA-targeted probes specific to other species within the *Enterococcus* group.

The first section of this method describes procedures for the synthesis and labeling of the Enterococci gene probes. Methods employed to examine environmental water samples are described, with details of specific hybridization and detection methodologies. Given the application of the *Enterococcus* group as microbiological indicators of water quality and the relevance of particular

Table 1
Enterococci Oligonucleotide Gene Probes

Probe	Target organism	Sequence[a]	T_m(°C)[b]	Target[c]
Efs (DB8)[a]	*Ent. faecalis*	5'-GGT GTT GTT AGC ATT TCG-3'	52	23S: 344–361
Efm (DB6)[c]	*Ent. faecium*	5'-CAC ACA ATC GTA ACA TCC-3'	52	23S: 141–158
Ent (Enc131)[a]	Most Enterococci	5'-CCC CTT CTG ATG GGC AGG-3'	60	16S: 131–147
EuB	Eubacterial	5'-GCT GCC TCC CGT AGG AGT-3'	60	16S: 338–355

[a] The *Ent. faecalis* and *Ent. faecium* sequences are based on those originally described by Betzl et al. ***(21)*** and later modified by Beimfohr et al. ***(30)***. The Enterococci probe (Enc131) was described by Meier et al. ***(26)***. The Eubacterial probe is that described by Amann et al. ***(20)***.

[b] T_m: Add 4°C for each G and C, 2°C for each A and T. Hybridize at 10°C below estimated T_m.

[c] Position in *E. coli* 16S rRNA primary structure ***(33)***.

Enterococcus species as fecal pollution indicators, a rapid direct lysis dot-blot screening protocol is also described to determine enterococci species identity. In brief, this isolate screening protocol involves the application of cell lysates onto a positively charged nylon membrane using a dot-blot apparatus using a modification of the method described by Byers et al. (1997) ***(31).*** Following alkali fixation, blots are hybridized with the enterococci oligonucleotide probes. Using this method the identity of enterococci isolates from environmental samples can be determined following the initial selective isolation and culture using the standard selective media recommended for water quality assessment ***(26,32)***.

2. Materials

2.1. Preparation of 23S and 16S rRNA Enterococci Oligonucleotide Gene Probes

1. Enterococci oligonucleotide probes (**Table 1**) synthesized to 50 nmol scale (Life Technologies) and stocks prepared at 20 pmol/μL in TE buffer yH8.0.
2. Probe labeling: Use DIG Oligonucleotide 3'-End Labeling Kit supplied by Roche Molecular Biochemicals, cat. no. 136 2372.

2.2. Analysis of Marine and Freshwater Samples

1. Positively charged nylon membrane (Hybond N+, Amersham), precut into 47-mm diameter circles.
2. Filtration apparatus: Swinnex-47 filter holder (Millipore), 50-mL syringe (Monoject) or equivalent filtration system.
3. Nonselective culturing media: Luria-Bertani (LB) agar: 1% (w/v) Bacto tryptone, 0.5% (w/v) Bacto yeast extract, 0.5% (w/v) NaCl, 1.5% (w/v) agar, adjust pH to 7.2.

4. BHI agar (Difco): 12.5 g of calf brain infusion solids, 5.0 g of beef heart infusion solids, 10.0 g of proteose peptone, 5.0 g of sodium chloride, 2.0 g of glucose, 2.5 g of disodium phosphate (anhydrous), 15.0 g of agar.
5. Filter blotting paper (Whatman 3MM).
6. Cell lysis solution (freshly prepared): 6 mg/mL of lysozyme (Roche) in 0.01 *M* Tris-HCl, pH 7.0.
7. Denaturing solution: 1.5 *M* NaCl; 0.5 *M* NaOH, prepared from sterile stock solutions.
8. Neutralizing solution: 1.5 *M* NaCl, 0.5 *M* Tris-HCl, pH 7.2, 0.001 *M* EDTA, pH 8.0; prepared from sterile stock solutions.
9. Alkai fixation solution: 0.4 *M* NaOH.
10. 5× saline sodium citrate (SSC): Prepared from 20× SSC stock. Per liter dissolve 175.3 g of NaCl; 88.2 g of sodium citrate. Adjust pH to 7.0 with 1 *M* HCl. Dispense in aliquots. Autoclave at 121°C, 20 min.

2.3. Enterococci Species Identification

1. Positively charged nylon membrane (Hybond N+, Amersham), precut membrane to fit a dot-blot manifold
2. Dot-blot apparatus: 96-Well dot-blot manifold (Bio-Rad, Bio-dot manifold).
3. Culture media: BHI broth (Difco).
4. Cell resuspension solution (TNE solution): 10 m*M* Tris-HCl, pH 8.0, 10 m*M* NaCl; 10 mM EDTA, pH 8.0. Prepare from sterile stock solutions.
5. Cell lysis solution: TNE solution + 2% Triton X-100 (Sigma).
6. Lysozyme stock solution: Prepare 50 mg/mL of lysozyme (Roche) in water. Dispense into aliquots and store at –20°C. Lysozyme stocks should be used only once following thawing.
7. 25% Sodium dodecyl sulfate (SDS): Dissolve 2.5 g of SDS in 90 mL of sterile water. Heat to 98°C to dissolve. Adjust the pH to 7.2 by adding a few drops of concentrated HCl; adjust to a final volume of 100 mL of with sterile water.
8. Filter blotting paper (Whatman 3MM).
9. Denaturing solution: 1.5 *M* NaCl; 0.5 *M* NaOH, prepare from sterile stock solutions.
10. Neutralizing solution: 1.5 *M* NaCl, 0.5 *M* Tris-HCl, pH 7.2, 0.001 *M* EDTA, pH 8.0; prepare from sterile stock solutions.
11. Alkali fixation solution: 0.4 *M* NaOH.
12. 5× SSC: Prepare 20× SSC stock. Per liter dissolve 175.3 g of NaCl, 88.2 g of sodium citrate. Adjust pH to 7.0 with 1 *M* HCl. Dispense in aliquots. Autoclave at 121°C, 20 min.

2.4. Hybridization and Detection

2.4.1. Hybridization

1. Prehybridization/hybridization solution: Per liter 5× SSC; 1% (w/v) blocking reagent (Roche), 0.1% (w/v) *N*-laurylsarcosine, Na salt; 0.02% (w/v) SDS. The

hybridization solution can be stored frozen at –20°C. Dissolve by heating at 50–70°C.

2. Hybridization bags (Life Technologies).
3. Post hybridization washes: Per liter: 2× SSC, 0.1% (w/v) SDS; Per liter: 0.1× SSC, 0.1% (w/v) SDS.
4. 10% SDS: Per 100 mL dissolve 1.0 g of SDS in 90 mL of H_2O. Heat to 68°C to dissolve. Adjust the pH to 7.0 with HCl.

2.4.2. Immunological Detection

Use the DIG Nucleic Acid Detection Kit supplied by Roche Molecular Biochemicals, cat. no. 1175 041.

3. Methods

3.1. Synthesis of 16S and 23S rRNA Enterococci Oligonucleotide Gene Probes

1. All oligonucleotide probes were synthesized by Life Technologies at a 50-nmol scale.
2. Resuspend probes in 50 µL of TE buffer, pH 8.0. Prepare stocks for labeling reactions to a final concentration of 20 pmol/µL. Enterococci oligonucleotide probe sequences are shown in **Table 1**.

3.1.1. Oligonucleotide Gene Probe Labeling

Label Enterococci oligonucleotide gene probes by incorporating a single DIG residue (DIG-11-ddUTP) (Roche) at the 3' end of the oligonucletide. Perform the 3'-end labeling reaction according to the manufacturer's instructions (DIG Oligonucleotide 3'-End Labeling Kit, Roche Molecular Biochemicals, cat. no. 1362372, http://biochem.roche.com) (*see* **Note 1**).

3.2. Analysis of Marine and Freshwater Samples Using the Enterococci Oligonucleotide Gene Probes

1. Filter appropriate volumes of fresh or marine water samples through 47-mm diameter (precut) positively charged nylon membrane, held in a Swinnex-47 (Millipore) filter holder or equivalent filtration device (*see* **Note 2**).
2. Place the membrane aseptically on brain heart infusion (BHI) agar and incubate in a sealed container for 15 h at 37°C (*see* **Note 3**).
3. Place the membranes on filter paper (Whatman 3MM) soaked in freshly prepared lysis solution to lyse the cells and release the RNA and DNA. Stack the membranes and incubate at 37°C for 1 h (*see* **Note 4**).
4. Following lysis blot the membranes, then place on filter paper soaked in denaturing solution for 7 min, blot dry.
5. Place the membranes on filter paper soaked in neutralizing solution for 5 min, blot dry.

6. To fix the nucleic acids to the membranes, place on filter paper soaked in 0.4 *M* NaOH (alkali fixation) for 30 min; blot dry.
7. Rinse the membranes with 5× SSC and gentle agitation for 1 min (*see* **Note 5**).
8. Air-dry the membranes and store at 4°C within sealed plastic bags (*see* **Note 6**).

3.3. Identification of Enterococcus Species by Dot-Blot Screening

3.3.1. Isolate Preparation

1. Grow enterococci isolates at 37°C overnight in 3 mL of BHI broth on a rotary shaker (200 rpm) (*see* **Note 7**).
2. Following incubation, harvest 1 mL of culture by centrifugation (14,000*g* for 1 min).
3. Wash bacterial cells in 1 mL of TNE solution and then resuspend cells in 196 µL of TNE solution.
4. Treat the cell resuspension with 4 µL of lysozyme (50 mg/mL solution) and incubate at 37°C for 30 min.
5. Add 22.2 µL of 25% SDS (to give a final concentration of 2.5% SDS in solution).
6. Incubate the tubes at room temperature for 20–30 min.

3.3.2. Manifold Preparation

1. Wash the dot-blot manifold (Bio-Rad, Bio-dot manifold) thoroughly with sterile water and air dry.
2. Place precut membrane on the manifold and assemble the unit. Ensure that the membrane is dry to prevent diffusion of the cellular debris.
3. Apply suction to the dot blot apparatus. As suction is applied tighten the manifold to ensure a good seal.
4. Apply 25 µL of the prepared cell suspensions and continue to apply suction for a further 2 min (*see* **Notes 8** and **9**).
5. Carefully dismantle the dot blot apparatus and allow the membrane to air-dry.

3.3.3. Nucleic Acid Fixation

1. Alkali fix nucleic acids to the membrane following the procedure described in **Subheading 3.2**, **steps 6–8**.

3.3.3.1 Enterococci Oligonucleotide Gene Probe Hybridization

1. Prehybridize membranes for 2 h at the probe-specific temperature in the prehybridization solution.
2. Replace the prehybridization solution with hybridization solution containing 1–10 pmol/mL of oligonucleotide probe. Seal membranes in a hybridization bag and incubate at the probe-specific temperature for 1–6 h (or overnight) with constant agitation (*see* **Notes 10** and **11**).
3. Following hybridization wash the membranes (2 × 5 min) in 2× SSC, 0.1% (w/v) SDS at the probe specific temperature (use approx 50 mL/100 cm^2 filter).

Membrane setup

*- Detection controls are indicated in **bold** type*

1	**2**	**3**	**4**	5
6	**7**	8	9	10
11	12	13	14	15
16	17	18	19	20

Key to isolate identity:

1. *Ent. faecalis* (ATCC[a] 19433) *(positive control)*
2. *Ent. faecium* (ATCC 19434) *(positive control)*
3. *Ent. faecalis* (NCTC[b] 5957) *(positive control)*
4. *E. coli* (ATCC 25922) *(negative control)*
5. *Ent. casseliflavus (environmental –marine water)*
6. *Ent. faecalis (environmental- marine water)*
7. Blank
8. *Ent. faecium (environmental – marine water)*
9. *Enterococcus spp. (environmental – marine water)*
10. *Ent. faecalis (environmental – marine water)*
11. *Staphylococcus aureus* (ATCC 25923) *(negative control)*
12. *Ent. durans (raw sewage)*
13. *Ent. faecium (environmental - freshwater)*
14. *Ent. faecium (environmental – marine water)*
15. *Ent. faecium (environmental – sand)*
16. *Ent. faecium (environmental – marine water)*
17. *Ent. faecalis (environmental – sand)*
18. *Ent. hirae (environmental – sediment)*
19. *Ent. faecium (environmental – marine water)*
20. *Ent. faecium (environmental – marine water)*

[a]ATCC – American Type Culture Collection, Rockville, Maryland, USA [b]NCTC – National Collection of Type Cultures, Central Public Health Laboratory, Colindale, London, England

Screening result

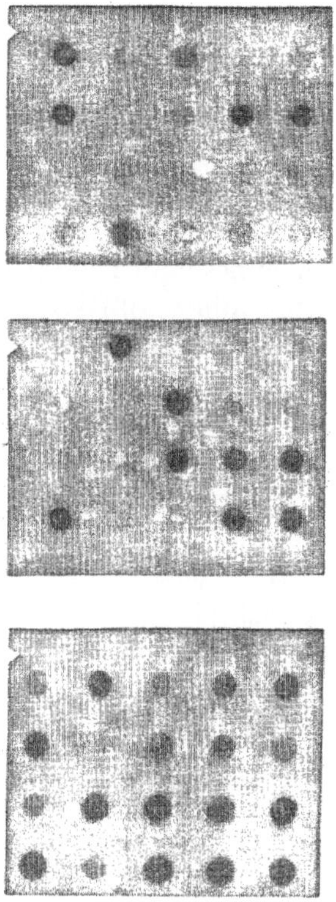

Probe

Efs (*Ent. faecalis*)

Efm (*Ent. faecium*)

EUB (Eubacterial)

Fig. 1. To demonstrate the utility of the dot-blot procedure for the identification of enterococci a series of isolates of known identity from a range of sources were screened against Efs (*Ent. Faecium*), Efs (*Ent. faecalis*), and EUB (*Eubacterial*) oligonucleotide gene probes.

4. Follow with washes of 2 × 5 min in 0.1× SSC, 0.1% (w/v) SDS at the probe specific temperature. Perform all washes with constant agitation.
5. Air-dry the membranes and store at 4°C for later detection (*see* **Notes 12** and **13**).

3.4. Immunological Detection

All hybridized membranes can be detected according to the manufacturer's instructions contained with the DIG Nucleic Acid Detection Kit (Roche Molecular Biochemicals, DIG Nucleic Acid Detection Kit, cat. no. 1175 041, http://biochem.roche.com). Detect the hybridized probes with antidigoxygenin-AP, Fab fragments (Roche) and visualize with the colourimetric substrates BCIP and NBT (Roche). A blue precipitate forms after a few minutes and is usually complete after approx 16 h; *see* **Fig. 1**. Wash the membranes in TE buffer to stop the detection reaction (*see* **Note 14**).

4. Notes

1. It is advisable to confirm the integrity and concentration of the labeled oligonucletides by following the procedures outlined in the DIG 3' end labeling kit (Roche).
2. An appropriate series of sample volumes should be selected such that a countable number (30–300) colonies will be present on the membrane filter. For marine and freshwaters triplicate volumes of 1.0 mL, 10 mL, and 100 mL volumes are generally sufficient except in instances where water quality may be compromised.
3. Mark the membranes with a soft pencil for later identification.
4. To minimize the volume of lysis solution stack the membranes, pipet 0.5 mL of lysis solution between each membrane.
5. Use the 5× SSC wash to remove cellular debris from the membrane surface. We have observed occasions where some nonspecific coloration occurs following detection of control isolates caused by a failure to remove all cellular debris.
6. Membranes have been stored for up to 2 mo at 4°C prior to hybridization.
7. Selection of isolates for this screening procedure requires that the isolates have been presumptively identified as belonging to the *Enterococcus* group. This identity would normally be obtained in the course of standard monitoring practice where primary isolation on selective culture media (mE/EIA ***[4]***, mEI ***[12]***) and confirmation of isolates (e.g., Gram-positive, growth in 6.5% salt, growth at 45°C, esculin-positive, and catalase-negative) will identify organisms as *Enterococcus*.
8. Using a 96-well dot-blot apparatus we were able to prepare four replicate membranes of 4 × 6 wells for multiple probe screening. Using this approach, 24 isolates could be screened against the three probes described without requiring membranes to be stripped and reprobed.
9. Positive (*Ent. faecalis* and *Ent. faecium* type strains) and negative (*E. coli* type strain) detection controls should be included within each screening series to ensure adequate cell lysis and detection specificity.
10. To minimise hybridization solution volumes: (a) Stack the membranes in the hybridization bag; between each membrane pipet 0.5 mL of the hybridization solution. (b) When heat sealing the hybridization bag, seal close to the membranes.
11. Hybridization solutions should be retained and can be reused several times. Store the used hybridization solutions at –20°C.
12. Hybridized membranes have been stored at 4°C for several weeks before detecting.
13. In the protocol described for detecting enterococci using the DIG-labeled oligonucleotide gene probes only one probe can be used at a time in any hybridization reaction. Membranes can be rehybridized following removal of the color substrate using dimethylformamide. A description of this procedure is contained in the manufacturer's instructions for the DIG Nucleic Acid Detection Kit (Roche Molecular Biochemicals). Membranes must not be allowed to dry if they are to be reprobed.

14. To minimize color substrate volumes, stack the membranes between layers of filter paper (Whatman 3MM) saturated in the color substrate solution. We have found that stacks of up to 10 membranes are developed efficiently. Take care to ensure that there are no air bubbles between the filter paper and membranes.

References

1. Barcina, I., Gonzalez, J. M., Iriberri, J., and Ega, L. (1990) Survival strategy of *Escherichia coli* and *Enterococcus faecalis* in illuminated fresh and marine systems. *J. Appl. Bacteriol.* **68,** 189–198.
2. Davies, C. M. and Evison, L. M. (1991) Sunlight and the survival of enteric bacteria in natural waters. *J. Appl. Bacteriol.* **70,** 265–274.
3. Evison, L. M. (1988) Comparative studies on the survival of indicator organisms and pathogens in fresh and sea water. *Water Sci. Technol.* **20,** 309–315.
4. Cornax, R., Morinigo, M. A., Romero, P., and Borrego, J. J. (1990) Survival of pathogenic microorganisms in seawater. *Curr. Microbiol.* **20,** 293–298.
5. Sinton, L. W., Davies-Colley, R. J., and Bell, R. G. (1994) Inactivation of enterococci and faecal coliforms from sewage and meatworks effluents in seawater chambers. *Appl. Environm. Microbiol.* **60,** 2040–2048.
6. Hackney, C. R. (1994) Methods to recover injured indicators and pathogens. In *Environmental Indicators and Shellfish Safety* (Hackney, C. R. and Pierson, M. D., eds.), Chapman and Hall, New York, pp. 218–273.
7. Colwell, R. R., Brayton, P. R., Grimes, D. J., Roszak, D. B., Huq, S. A., and Palmer, L. M. (1985) Viable but non-culturabe *Vibrio cholerae* and related pathogens in the environment: implications for release of genetically engineered microorganisms. *BioTechnology* **3,** 817–820.
8. Roszak, D. B. and Colwell, R. R. (1987) Survival strategies of bacteria in the natural environment. *Microbiol. Rev.* **51,** 365–379.
9. Xu, H., Roberts, N., Singleton, F. L., Atwell, R. W., Grimes, D. J., and Colwell, R. R. (1982) Survival and viability of nonculturable *Escherichia coli* and *Vibrio cholerae* in the estuarine and marine environment. *Microb. Ecol.* **8,** 313–323.
10. Garcia-Lara, J., Menon, P., Servais, P., and Billen, G. (1991) Mortality of faecal bacteria in seawater. *Appl. Environm. Microbiol.* **57,** 885–888.
11. Grimes, D. J., Atwell, R. W., Brayton, P. R., Palmer, L. M., Rollins, D. M., Roszak, D. B., et al. (1986) The fate of enteric bacteria in estuarine and marine environments. *Microbiolog. Sci.* **3,** 324–329.
12. Roszak, D. B., Grimes, D. J., and Colwell, R. R. (1984) Viable but nonrecoverable stage of *Salmonella enteritidis* in aquatic systems. *Can. J. Microbiol.* **30,** 334–338.
13. Grimes, D. J. and Colwell, R. R. (1986) Viability and virulence of *Escherichia coli* suspended by membrane chamber in semitropical ocean water. *FEMS Microbiol. Lett.* **34,** 161–165.
14. Pommepuy, M., Butin, M., Derrien, A., Gourmelon, M., Colwell, R. R., and Cormier, M. (1996a) Retention of enteropathogenicity by viable but nonculturable

Escherichia coli exposed to seawater and sunlight. *Appl. Environm. Microbiol.* **62,** 4621–4626.
15. McFeters, G. A., Cameron S. C., and Chevallier, M. W. L.. (1982) Influence of diluents, media, and membrane filters on detection of injured waterborne coliform bacteria. *Appl. Environm. Microbiol.* **43,** 97–103.
16. Lleo, M. M., Tafi, M. C., and Canepari, P. (1998) Nonculturable *Enterococcus faecalis* cells are metabolically active and capable of resuming active growth. *System. Appl. Microbiol.* **21,** 333–339.
17. Evison, L. M. and Tosti, E. (1981) An appraisal of bacterial indicators of pollution in sea water. *Water Sci. Technol.* **13,** 591–599.
18. Gameson, A. L. H. and Gould, D. J. (1974) Effects of solar radiation on the mortality of some terrestrial bacteria in sea water.Presented at the Discharge of sewage from sea outfalls. 27 August to 2 September 1974. Gameson, A. L., ed., 209–219, Pergamon Press, New York.
19. LeChevallier, M. W. and McFeters, G. A. (1985) Enumerating injured coliforms in drinking water. *J. Am. Water Works Assoc.*(**June**), 81–87.
20. Amann, R. I., Krumholz, L., and Stahl, D. A. (1990b) Fluorescent-oligonucleotide probing of whole cells for determinative, phylogenetic, and environmental studies in microbiology. *J. Bacteriol.* **172,** 762–770.
21. Betzl, D., Ludwig, W., and Schleifer, K. H. (1990) Identification of lactococci and enterococci by colony hybridization with 23S rRNA-targeted oligonucleotide probes. *Appl. Environm. Microbiol.* **56,** 2927–2929.
22. Frahm, E., Heiber, I., Hoffmann, S., Koob, C., Meier, H., Ludwig, W., et al. (1998) Application of 23S rDNA-targeted oligonucleotide probes specific for Enterococci to water hygiene control. *System. Appl. Microbiol.* **21,** 450–453.
23. Meier, H., Koob, C., Ludwig, W., Amann, R., Frahm, E., Hoffmann, S., et al. (1997) Detection of enterococci with rRNA targeted DNA probes and their use for hygienic drinking water control. *Water Sci. Technol.* **35,** 437–444.
24. Green, D. H., Lewis, G. D., Rodtong, S., and Loutit, M. W. (1991) Detection of faecal pollution by an *Escherichia coli uidA* gene probe. *J. Microbiolog. Meth.* **13,** 207–214.
25. Anderson, S. A., Lewis, G. D., and Pearson, M. N. (1995) Use of gene probes for the detection of quiescent enteric bacteria in marine and fresh waters. *Water Sci. Technol.* **31,** 291–298.
26. APHA. (1992) Standard methods for the examination of water and wastewater. In 18th ed. American Public Health Association, American Water Works Association, Water Environment Federation, Washington, DC.
27. Lewis, D. L. and Gattie, D. K. (1991) The ecology of quiescent microbes. *ASM News* **57**, 27–32.
28. Amann, R. I., Binder, B. J., Olsen, R. J., Chisholm, S. W., Deveraux, R., and Stahl, D. A.. (1990a) Combination of 16S rRNA-targeted oligonucleotide probes with flow cytometry for analysing mixed microbial populations. *Appl. Environm. Microbiol.* **56,** 1919–1925.
29. Stahl, D. A. and Amann, R. (1991) Development and application of nucleic acid

probes. In *Nucleic Acid Techniques in Bacterial Systematics* (Stackebrandt, E. and Goodfellow, M., eds.), John Wiley & Sons, Chichester, New York, pp. 205–248.

30. Beimfohr, C., Krause, A., Amann, R., Ludwig, W., and Schleifer, K. (1993) *In situ* identification of Lactococci, Enterococci and Streptococci. *System. Appl. Microbiol.* **16,** 450–456.
31. Byers, H. K., Patel, B., and Stackebrandt, E. (1997) Design and application of four oligonucleotide probes specific for *Thermus* species. *System. Appl. Microbiol.* **20,** 248–254.
32. EPA. (1997) Method 1600: membrane filter test method for enterococci in water EPA-821-R-97-004. United States Environmental Protection Agency.
33. Brosius, J., Dull, T. J., Sleeter, D. D., and Noller, H. F. (1981). Gene organization and primary structure of a ribosomal RNA operon from *Escherichia coli. J. Mol. Biol.* **148,** 107–127.

16

Analysis of 16S–23S rRNA Intergenic Spacer of *Vibrio cholerae* and *Vibrio mimicus* for Detection of These Species

Jongsik Chun, Irma N. G. Rivera, and Rita R. Colwell

1. Introduction

Vibrio cholerae, a noninvasive, Gram-negative bacterium responsible for severe epidemics of cholera and endemic diarrhea in many parts of the world, especially in developing countries, is a native inhabitant of brackish and estuarine ecosystems ***(1,2)***. Of approx 193 serogroups of *V. cholerae*, serogroups O1 and O139 are the causative agents of cholera epidemics. However, serogroups that are non-O1/non-O139 have been associated with small outbreaks of diarrheal disease and have been isolated from patients with intestinal or extraintestinal infections. Molecular detection of the pathogenic serogroups has been accomplished using polymerase chain reaction (PCR) or nucleic acid probe assays based on genes coding for cholera toxin (*ctxAB*) ***(3)***. This method has limitations, namely, that it detects all *ctxAB* containing strains, regardless of taxonomic identity. For example, *Vibrio mimicus* and *Aeromonas* spp. strains have been shown to possess *ctxAB* genes. In addition, the method may fail to detect potentially pathogenic *V. cholerae* strains that do not contain *ctxAB* genes. It has been shown that the *ctxAB* gene can be transferred between *V. cholerae* strains via filamentous phages ***(4)***. Therefore, it is important to detect *V. cholerae* strains at the species level. Unfortunately, methods based on 16S or 23S rDNA sequences are not useful because sequences for *V. cholerae* and closely related *V. mimicus* are almost identical. Chun et al. ***(5)*** determined sequences of 16S–23S rRNA intergenic spacer regions of *V. cholerae* and *V. mimicus*, and found a region suitable for differentiating *V. cholerae* strains from other bacteria, including *V. mimicus*. Using a pair of oligonucleotide prim-

From: *Methods in Molecular Biology, vol. 179: Gene Probes: Principles and Protocols*
Edited by: M. Aquino de Muro and R. Rapley © Humana Press Inc., Totowa, NJ

Fig. 1. Identification of *V. cholerae* using PCR based on the 16S–23S rRNA intergenic spacer region. Lanes M, molecular weight marker (100 bp ladder): *1*, *V. cholerae* O1 classical ATCC 14035^{T}; *2–5*, *V. cholerae* O1 El Tor clinical isolates; *6–9*, *V. cholerae* O139 clinical isolates; *10*, *V. mimicus* ATCC 33653 ^{T}; *11–15*, *V. mimicus* isolates; *16–18*, *V. cholerae* non-O1/non-O139 isolates; *19*, *V. aestuarianus* ATCC35048^{T}; *20*, *V. alginolyticus* ATCC17749^{T}; *21*, *V. campbellii* ATCC25920^{T}; *22*, *V. carchariae* ATCC35084^{T}; *23*, *V. diazotrophicus* ATCC33466^{T}; *24*, *V. fischeri* ATCC7744^{T}; *25*, *V. fluvialis* ATCC33809^{T}; *26*, *V. furnissii* ATCC35016^{T}; *27*, *V. hollisae* ATCC33564^{T}; *28*, *V. natriegens* ATCC14048^{T}; *29*, *V. salmonicida* ATCC43839^{T}; *30*, *V. vulnificus* ATCC27562^{T}. ^{T} Type strain.

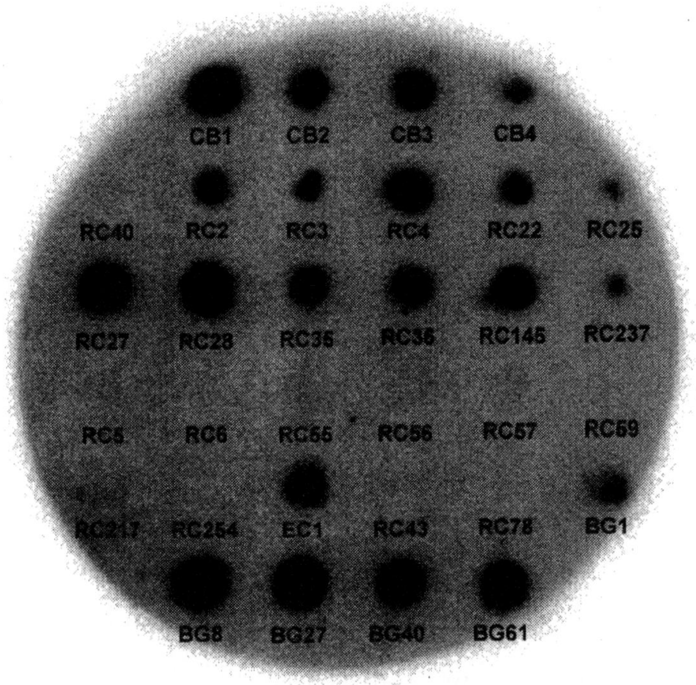

Fig. 2. An example of *V. cholerae* identification using colony hybridization. Blots: CB1–CB4, *V. cholerae* isolates from Chesapeake Bay; BG1–BG61, EC1, RC2–RC4, RC22–RC36, RC145, and RC237, *V. cholerae* isolates from different countries of the world; RC5–RC6, RC55–RC59, RC217, and RC254, *V. mimicus*; RC40, *Aeromonas jandaei*; and RC43 and RC78, *V. vulnificus*.

ers, the PCR assay yields 300-basepair (bp) amplicons highly specific for *V. cholerae* strains (**Fig. 1**). For high-throughput identification, colony hybridization can be used for counting culturable cells (**Fig. 2**). This chapter describes a detailed procedure for PCR and colony hybridization to detect *V. cholerae* in environmental samples and/or its identification in isolated cultures.

2. Materials

2.1. Culture of V. cholerae and Related Bacteria

1. Luria–Bertani (LB) medium (Difco), containing 2% NaCl (total concentration).

2.2. Extraction of Chromosomal DNA

1. 50 mg/mL of lysozyme in TE buffer.
2. Guanidine-sarkosyl solution. Add 60.0 g of guanidine thiocyanate (Sigma), 20 mL of EDTA (0.5 m*M*, pH 8.0), and 20 mL of Milli-Q water to a sterile bottle.

Heat at 65°C until dissolved. After cooling to room temperature, add 5 mL of 10% *N*-lauroylsarcosine, sodium salt (sarkosyl) (Sigma). Complete the volume to 100 mL with Milli-Q water, filter through a 0.45-µm filter (Millipore), and store at room temperature.

3. 7.5 *M* Ammonium acetate.
4. Chloroform–Isoamyl alcohol (24:1).
5. 2-Propanol (isopropanol).
6. TE buffer or Milli-Q sterile water.

2.3. V. cholerae-*Specific PCR*

1. 10× PCR buffer (Promega).
2. dNTP mix solution: 2.5 m*M* each of dATP, dCTP, dGTP, dTTP (Promega).
3. *Taq* DNA polymerase (Promega; 5 U/mL).
4. Primers to amplify 16S–23S rRNA spacer region
   pVC-F (5'—TTA AGC [C/G]TT TTC [A/G]CT GAG AAT G-3')
   pVCM-R (5'—AGT CAC TTA ACC ATA CAA CCC G-3')
   Primer stock solutions are prepared to a final concentration of 20 µ*M*.
5. 50 ng of genomic DNA or 5 mL of crude DNA extract (*see* **Subheading 3.2.2.**).
6. Loading dye II_m: 0.05% Bromophenol blue, 0.05% xylene cyanole FF, and 15% Ficoll 400 in water (store at room temperature) (modified from **ref. 7**).

2.4. Colony Hybridization

2.4.1. Radioactive Labeling of pVC-F

1. Oligonucleotide pVC ITS-1 (20 µM):
   pVC ITS-1 (5'-GC [C/G]TT TTC [A/G]CT GAG AAT G-3').
2. 10× T4 polynucleotide kinase buffer.
3. [γ-^{32}P]ATP (20 pmol).
4. T4 Polynucleotide kinase.
5. Milli-Q sterile water.
6. QIAquick Nucleotide Removal Kit.

2.4.2. Preparation of Colony Blots

1. Luria-Bertani agar (2% NaCl) plates.
2. Toothpicks.
3. Whatman 541 filter paper.
4. Whatman 3MM filter paper.
5. Alkaline buffer: 0.5 *M* NaOH, 1.5 *M* NaCl.
6. Neutralization buffer: 2 M ammonium acetate.
7. 1× Saline sodium citrate (SSC) solution.

2.4.3. Prehybridization and Hybridization

1. 3× SSC/0.1% sodium dodecyl sulfate (SDS).

2. Prehybridization buffer: 6× SSC, 5× Denhardt's solution, 0.05% sodium pyrophosphate, 100 µg/mL of boiled salmon sperm DNA, 0.5% SDS.
3. Hybridization buffer: 6× SSC, 1× Denhardt's solution, 0.05% sodium pyrophosphate.
4. 6× SSC/0.05% sodium pyrophosphate.

3. Method

3.1. V. cholerae *and Related Bacterial Culture*

Inoculate *V. cholerae* or another strain at 30°C in LB broth or agar (2% NaCl). For colony hybridization, strains are inoculated on LB agar plates (90 mm), using sterile toothpicks to transfer, and grown overnight at 30°C.

3.2. Extraction of Chromosomal DNA

For PCR, chromosomal DNA can be extracted using two different methods.

*3.2.1. DNA Isolation Using Guanidine Thiocyanate (Modified After **ref. 8**)*

1. Centrifuge overnight cultures at 12,000 rpm for 2 min. Alternatively, a loopful of biomass can be scraped from the surface of LB plates.
2. Resuspend a small (rice-grain size) cell pellet in 100 µL of fresh lysozyme (50 mg/mL) in TE buffer, and incubate at 37°C for 30 min. Lysozyme may be omitted.
3. Lyse the cells with 500 µL of guanidine–sarcosyl solution.
4. Vortex-mix the microcentrifuge tubes briefly and check for lysis (5–10 min).
5. Cool the lysates on ice, add 250 µL of cold 7.5 *M* ammonium acetate and mix. Hold on ice for an additional 10 min.
6. Add 500 µL of chloroform–isoamyl alcohol (24:1), mix thoroughly, and centrifuge at 13,000 rpm for 10 min.
7. Transfer the supernatant to a 1.5-mL microcentrifuge tube.
8. Add 0.6 volume of 2-propanol (isopropanol).
9. Homogenize the tubes for 1 min to mix and centrifuge the DNA at 13,000 rpm for 1–10 min.
10. Wash the pellet three times with 70% ethanol solution and dry under vacuum.
11. Dissolve the DNA in 50–100 µL of TE buffer or Milli-Q sterile water and store at 4°C.
12. DNA concentration is determined by measuring A_{260} using a UV spectrophotometer.
13. Prepare DNA stock with a concentration of 10 ng/µL for PCR.

3.2.2. Preparation of Crude DNA Extract

Crude DNA extract can be prepared by boiling cells.

1. Transfer a *V. cholerae*-like colony to a 1.5-mL microcentrifuge tube containing 500 µL of Milli-Q sterile water.
2. Boil for 10 min and cool to room temperature.
3. Centrifuge at 13,000 rpm briefly.
4. Dilute the supernatant in a clean tube 1:1000 with Milli-Q sterile water.
5. Measure the DNA concentration using a UV spectrophotometer (optional) (*see* **Note 3**).
6. Use a 5 µL aliquot in 10× TAE buffer for PCR.

3.3. V. cholerae-Specific PCR

1. Prepare the PCR master mix per sample as follows (*see* **Note 1**):
   5 µL of 10× PCR buffer
   4 µL of dNTP mix (2.5 m*M*)
   2 µL of primer pVC-F stock (20 µ*M*)
   2 µL of primer pVCM-R stock (20 µ*M*)
   31.75 µL of sterile water
   0.25 µL of *Taq* polymerase (5 U/µL)
2. Add 45 µL of the PCR master mix to 0.5-mL microcentrifuge tubes, and overlay each with a drop of mineral oil.
3. Place tubes in a thermocycler and heat the cycler to 72°C.
4. Add 5 µL of the DNA sample to the preheated PCR mix.
5. Start the amplification with the following cycle parameters:

Initial denaturation:	94°C, 2 min
30 cycles of:	94°C, 1 min
	60°C, 1 min
	72°C, 1 min
Final extension:	72°C, 10 min

6. Mix 20 µL of the resultant PCR reactions with 4 µL of loading dye II_m and load onto a 1.5% agarose gel. Include a nucleic acid suitable molecular weight marker. Perform electrophoresis in 1X TAE buffer.
7. Stain gel in ethidium bromide solution and visualize bands under a UV illuminator.
8. Only *V. cholerae* strains produce a 300-bp band.

3.4. Colony Hybridization

3.4.1. Radioactive Labeling of pVC-F

1. Prepare the mix as follows:

10 µL	pVC ITS-1 (20 µ*M*)
4 µL	10× T4 polynucleotide kinase buffer
8 µL	[γ-^{32}P]ATP (sp act 5000 Ci/mmol; 10 mCi/mL in aqueous solution) = 20 pmol
17 µL	Milli-Q sterile water
1 µL	T4 polynucleotide kinase

2. Mix well and incubate for 30 min at 37°C.
3. Incubate the tube for 15 min at 65°C to inactivate the kinase.
4. Purify labeled oligonucleotide using a QIAquick Nucleotide Removal Kit (Qiagen Inc. CA).

*3.4.2. Preparation of Colony Blots (see **Note 2**)*

1. Inoculate bacteria using sterile toothpicks onto LB agar plates and incubate the inoculated plates at 30°C for 16–24 h.
2. Prepare colony blots by overlaying filter papers (Whatman 541) onto the agar plates and leave for 1 h.
3. Place the filter paper carefully, bacterial side up, on Whatman 3MM paper saturated with alkaline buffer.
4. Microwave (500 W) the filter papers for 2 min.
5. Transfer the filters (face up) to neutralization buffer and allow to stand for 5 min at room temperature.
6. Wash the filters twice with 1× SSC solution.
7. Dry the filters. Store at –20°C, if not used immediately.

*3.4.3. Prehybridization and Hybridization in Sodium Chloride/Sodium Citrate (Based on **ref. 9**)*

1. Wash the filters in 3× SSC/0.1% SDS, three times, at room temperature and then in the same solution at 65°C for at least 2 h.
2. Incubate the filters in prehybridization buffer (10 mL per two filters) at 37°C for 1 h in a sterile bottle or sealable bag.
3. Remove the prehybridization buffer, add 10 mL of hybridization solution prewarmed to 42°C and add 40 μL of labeled pVC ITS-1 probe.
4. Perform hybridization at 42°C overnight (12–24 h).
5. Wash the filters 3×, each time for 15 min at room temperature in 6× SSC/0.05% sodium pyrophosphate.
6. Wash the filters in prewarmed 6× SSC/0.05% sodium pyrophosphate for 1 h at 55°C.
7. Seal the individual filters with plastic wrap, and expose to X-ray film for 16 h (overnight) at –70°C.

4. Notes

1. If 0.2-mL PCR tubes and an oil-free thermal cycler (e.g., Perkin Elmer 9600) are used for the PCR assay, the total volume can be reduced from 50 to 25 μL.
2. Primary isolation plates, such as alkaline peptone agar (APA) or R2A, can be used to enumerate *V. cholerae*. (Add 100 μL of seawater or other water sample on APA, spread, and incubate at 30°C for 24–72 h.)
3. All reagents must be molecular biology grade.
4. The best diluent for DNA is Milli-Q sterile water (because it does not interfere with the PCR reaction).

5. The amount of isolated DNA may vary, so, it is recommended to measure the DNA concentration before doing the PCR.

References

1. Colwell, R. R., Kaper, J., and Joseph, S. W. (1977). *Vibrio cholerae, Vibrio parahaemolyticus* and other *vibrios*: occurrence and distribution in Chesapeake Bay. *Science* **198,** 394–396.
2. Colwell, R. R. (1996) Global climate and infectious disease: the cholera paradigm. *Science* **274**, 2025–2031.
3. Rivera, I. G., Chowdhury, M. A. R., Sanchez, P. S., Sato, M. I., Huq, A., Colwell, R. R., and Martins, M. T. (1995) Detection of cholera (*ctx*) and zonula occludens (*zot*) toxin genes in *Vibrio cholerae* O1, O139 and non-O1 strains. *World J. Microbiol. Biotechnol.* **11,** 572–577.
4. Waldor, M. K. and Mekalanos, J. J. (1996) Lysogenic conversion by a filamentous phage encoding cholera toxin. *Science* **272,** 1910–1914.
5. Chun, J., Huq, A., and Colwell, R. R. (1999) Analysis of 16S-23S rRNA intergenic spacer regions of *Vibrio cholerae* and *Vibrio mimicus*. *Appl. Environ. Microbiol.* **65,** 2202–2208.
6. Sambrook, J., Fritsch, E. F., and Maniatis, T. (1989) *Molecular Cloning: A Laboratory Manual*, 2nd edit., Cold Spring Harbor Laboratory Press, Cold Spring Harbor, New York.
7. Pitcher, D. G., Saunders, N. A., and Owen, R. J. (1989) Rapid extraction of bacterial genomic DNA with guanidium thiocyanate. *Lett. Appl. Microbiol.* **8,** 151–156.
8. *Current Protocols in Molecular Biology* (1987). (Item 6.4.1). John Wiley & Sons, New York.

17

Colony Hybridization of Bacterial Isolates with *Burkholderia cepacia*-Specific Probes

Laura G. Leff

1. Introduction

Colony hybridization is a very powerful tool for the examination of bacterial isolates that have been cultured. The advantages are that it can be used rather rapidly with a high degree of specificity to look at features of large numbers of isolates ***(1)***. For *Burkholderia* (formerly *Pseudomonas*) *cepacia*, species-specific probes that target the 16S and 23S rDNA are available and can be easily used in colony hybridization (***2*** and ***3***, respectively). Researchers are interested in *B. cepacia* for several reasons: it is abundant in nature ***(4–6)***, it is valuable in bioremediation and biocontrol ***(7)***, and it is clinically important, causing disease in cystic fibrosis patients and other compromised individuals ***(8)***.

Although *B. cepacia* is important and widely studied, one difficulty is the rather complex and unresolved taxonomy of this species. The group of related strains and species is often referred to as the *B. cepacia* complex. The species is divided up into several genomovars, some of which have recently been designated as new species ***(9)***. Prior to the division of the complex into genomovars, it was divided into subgroups based on fatty acid composition ***(10)***. It has not been established if the probes described below are effective against all genomovars or if they will hybridize with other species in the genus *Burkholderia*. As the taxonomy of this complex is fully established, the breadth and specificity of the method will become more clear. Nevertheless, the procedure provided here can be used with a variety of probes and is easily modified to accommodate different probes. When this modification is attempted several factors must be considered: (1) obtaining and labeling the probe (which

From: *Methods in Molecular Biology, vol. 179: Gene Probes: Principles and Protocols*
Edited by: M. Aquino de Muro and R. Rapley © Humana Press Inc., Totowa, NJ

depends on the size and target), (2) determining stringency of conditions (modifying temperature and buffers to get appropriate stringency), and (3) running adequate positive and negative controls (selected based on the particular hybridization target). It is also noteworthy that the procedure below represents a method that does not rely on radioactive labeling, lessening safety concerns, regulations and expenses.

2. Materials

2.1. Blotting of Colonies

1. Blotting membrane, for example, Micron Separations, Inc. Magna Lift.
2. SSC: 20× Stock solution-dissolve 175.3 g of sodium chloride and 88.2 g of sodium citrate in 960 mL of deionized water, adjust pH to 7.0, and bring up to 1000 mL.
3. 0.5 *M* Tris-HCl, pH 8.0.
4. 95% Ethanol.
5. 0.5 *M* NaOH.

2.2. Labeling of Probes

1. DIG oligonucleotide 3' end labeling kit (Roche Molecular Biochemicals).
2. 0.5 *M* EDTA: pH must be adjusted to allow dissolution, while the solution is on a stir plate, adjust the pH to 8.0 with solid NaOH.
3. Oligonucleotide probe(s):16S rRNA for *B. cepacia* = 5'CCTCTGTTCCGACCA3' *(2)*; 23S rRNA for *B. cepacia* = 5'CCCATCGCATCTAACAAT3' *(3)*.

2.3. Purification of Probes

1. Probe purification push columns (Nuc-Trap®, Stratagene Inc., La Jolla, California).
2. STET: 100 m*M* NaCl, 20 m*M* Tris-HCl, pH 7.5, 10 m*M* EDTA, pH 8.0, 0.1% Tween-20.
3. 1% Tween-20.

2.4. Hybridization of Membranes

1. Buffer 1: Maleic acid buffer, pH 7.5: 0.10 *M* maleic acid, 0.15 *M* NaCl, pH adjusted to 7.5 with solid NaOH.
2. Buffer 2: Blocking solution: 10% blocking stock solution in buffer 1, pH 7.5, store at 4°C.
3. Blocking stock solution (10× concentration), store at 4°C: 10% blocking reagent (Roche Molecular Biochemicals, Indianapolis, Indiana) in buffer 1, Heat to 65°C with constant stirring, cool to 22°C. Adjust pH to 7.5.
4. Buffer 3: 0.1 *M* Tris-HCl, 0.1 *M* NaCl, pH 9.5.
5. Washing buffer: 0.3% v/v Tween in buffer 1, pH 7.5.
6. Hybridization buffer: 5× Saline sodium citrate (SSC), 0.5% blocking stock solu-

tion, 0.1% *N*-lauroylsarcosine, 0.02% sodium dodecyl sulfate (SDS).

7. 10% SDS—Be cautious and follow MSDS sheet when preparing (wear a mask to prevent dust inhalation and wipe down the area after weighing); heat to 68°C to dissolve and adjust pH to 7.2.
8. SSC (see **Subheading 2.1.**).
9. DIG chemilumescent detection kit, CSPD: disodium 3—(4-methoxyspiro {1,2-dioxetane-3, 2' (5'-chloro) tricyclo [3.3.1.1.3,7] decan}-4-4' phenyl phosphate; and anti-digoxigenin antibody-alkaline phosphatase conjugate (Roche Molecular Biochemicals).
10. Hybridization shaker or incubator (and heat-sealable bags, as appropriate for incubator style).
11. X-ray film and cartridges.
11. Chemicals for developing film: Kodak GBX fixer, developer and photoflo.

3. Methods

3.1. Blotting of Colonies

The bacteria of interest should be grown on appropriate media (for *B. cepacia*, nutrient agar often suffices). Bacterial colonies should be grown on solid media for 24–48 h and then blotted as described below. It is useful to arrange the bacterial colonies in numbered squares on the plates (*see* **Note 1**). Appropriate positive and negative controls should be included on each plate and each plate should be labeled with a short identification code. Sterile toothpicks can be used to quickly transfer colonies to the appropriate location on a plate. If the bacterial colonies are to be tested further (after the hybridization results are obtained), a duplicate plate should be made with the same colonies in the same order. The blotting protocol is based on the method of Shimkets and Asher ***(11)***.

1. Use circular pieces of positively charged nylon (such as Micron Separations, Magna Lift). Handle nylon with gloved hands so as not to contaminate your samples. Label one side of the filter, with pencil, with the identification code from the plate (write small and close to the edge).
2. Cut a distinctive notch or series of notches along the edge of each filter. This allows you to properly orient the film (that will show the hybridization results) with the plates. This is also helpful in identifying the images on the film; sometimes the pencil notation from the filter can be seen on the film but often it is not visible. The edge of the filter showing the notches is typically quite clear on the film.
3. Lay filters down on the plates (containing the bacterial colonies of interest). The filter should be placed with the identification code (written in pencil) facing down. Leave in place for 5 min. Be careful not to smear the colonies.
4. Mark the location and shape of the notches on the plates with a marker so the

filter and plate can be aligned in the proper orientation when the procedure is completed. Plates should be retained until the hybridization procedure is completed (**Subheading 3.4.**).

5. Remove filters from plates using forceps. Lift filters straight up to avoid smearing the colonies. Place on a large piece of filter paper with the side with the bacterial colonies (and the pencil notation) up.
6. Place filters (leave on the filter paper) in 37°C incubator for 15 min to dry the colonies.
7. Place filters in a sealable plastic container on a shaker at room temperature. To wash, add about 100 mL (for every five filters) of the appropriate solution to the plastic container containing the filters. Put on the lid of the container and shake for 2.5 min. Pour off the wash solution and continue as below. Repeat each wash two times.
   0.5 *M* NaOH for 2.5 min (twice)
   0.5 *M* Tris-HCl, pH 8.0 for 2.5 min (twice)
   2× SSC for 2.5 min (twice)
   Ethanol (95%) for 2.5 min (twice)
8. Bake at 80°C for 2 h (*see* **Note 2**). Filters can then be stored until use.

3.2. Labeling of Probes

The procedure below is for end labeling of oligonucleotide probes that have been synthesized. Other labeling methods can be used, especially in conjunction with larger probes. This method involves 3' end labeling of the probe with digoxigenin.

1. These ingredients can be purchased in a kit from Roche Molecular Biochemicals (*see* **note 4**). This makes enough for three hybridization procedures.
   Into a 0.5-mL microcentrifuge tube, mix
   - 21 µL sterile, deionized water
   - 12 µL terminal transferase buffer
   - 12 µL cobalt chloride ($CoCl_2$)
   - 300 pmoles oligonucleotide (rehydrated, if freeze dried; *see* **Note 3**)
   - 3 µL digoxigenin ddUTP
   - 3 µL terminal transferase
2. Incubate for 15 min at 37°C.
3. Add 3 µL of 0.5 *M* EDTA and freeze at –20°C. The probe should be purified before use.

3.3. Purification of Probes

This procedure removes excess label from the probe labeling mixture.

1. Add 5 µL of 1% Tween-20 and 5 µL of STET to the labeling reaction from section 3.2.
2. Prewet a probe purification push column by applying 70 µL of STET. Attach a

syringe to the column and push buffer through with a constant, even pressure. Discard any buffer exiting the column.
3. Apply probe solution to column and push it through using a syringe. Collect the column effluent in a 1.5-mL microcentrifuge tube.
4. Wash column with 70 μL of STET and collect effluent.
5. Measure the volume of the effluent. Use one third of the total volume per hybridization (each hybridization procedure can accommodate up to 10 filters). Freeze at –20°C until needed.

3.4. Hybridization of Membranes

This method is based on the protocol described in Leff et al. ***(1)***. A chemilumescent detection system is used in which the digoxigenin-labeled probes are detected using an immunoassay based system.

1. Set the temperature on the hybridization shaker/incubator (*see* **Note 5**); use a water bath if appropriate for the style of incubator. For the 16S rDNA probe for *B. cepacia* the temperature is 50°C and for the 23S rDNA probe it is 47°C.
2. Place filters in a plastic bag that can be heat sealed (or in the hybridization vessel appropriate for your style of incubator; *see* **Notes 5** and **6**). Ten or fewer filters should be placed in a single bag. If more than ten filters need to be hybridized, use multiple bags. To prehybridize: add hybridization buffer, seal bag, and place in incubator for 2 h. Use about 20 mL of hybridization buffer per 10 cm^2 of filters.
3. Cut open corner of bag and pour out buffer. Add 5 mL of hybridization buffer and labeled probe (use one third of the total labeled probe solution volume from **Subheading 3.3.**). Reseal bag and return to incubator. Hybridize overnight at correct temperature with shaking. Put a bottle of 1× SSC, 1% SDS in the shaker (make sure the top of the container is securely sealed) for use the next day (in **step 5**).
4. Open bag and pour out hybridization solution. Wash filters two times in 2× SSC, 1% SDS for 5 min each at room temperature with shaking (about 200 mL in each wash).
5. Wash filters 3× in 1× SSC, 1% SDS (this was placed in the incubator in **step 3**) for 20 min each at the hybridization temperature.
6. Wash with washing buffer for 1 min at room temperature.
7. Put in another heat sealable bag with 225 mL of buffer 2. Seal and shake for 30 min at room temperature.
8. Cut the bag open and remove buffer 2. Add 25 mL of buffer 2 and 5 μ anti-digoxigenin antibody–alkaline phosphatase conjugate. Seal and shake for 30 min at room temperature.
9. Wash 2× in washing buffer for 15 min each wash at room temperature.
10. Wash one time in buffer 3 for 2 min at room temperature.
11. Cut acetates (i.e., overhead transparencies) to fit into an X-ray film cartridge. Create pools consisting of 20 drops of CSPD on the transparency; one pool per

filter. Place each filter DNA side down (this is the side with the pencil notation) on a CSPD pool. Place another transparency on the back of the filters. Press together and wrap in plastic wrap; smooth out wrinkles.

12. Let stand 5 min at room temperature.
13. Incubate 15 min at 37°C.
14. Place in X-ray film cartridge with the pencil notations up and place a piece of film on top (*see* **Note 7**). Seal the cartridge and place in light-proof nylon bag (as needed). Expose film for 45–60 min at room temperature.
15. To develop film, put the film in developing trays in the order below for the times indicated. Fill each tray with 500 mL of the appropriate solution (*see* **Note 8**).
    5 min in developer (4:1 water–developer)
    1 min in tap water
    5 min in fixer (4:1 water–fixer)
    15 min in water bath
    30 s in Photoflo (1 capful–500 mL of H_20)
    Hang film to dry.
16. Film can be aligned correctly with the original plates containing the bacterial colonies using the notches visible on the film. To accomplish this alignment, lay the film over the plate and align the image of the notches on the film with the marked notches on the plates. In this way, the hybridization results can be determined for each specific bacterial colony. Bacterial colonies of interest can be recovered as needed from replicate plates.

4. Notes

1. To make grids on plates, markers or custom made rubber stamps can be used. Often, it is more efficient to lay the plate over a guide without actually marking the plates. Guides and plate holders can be made by hand or purchased (Easi-Grid, Jencons [Scientific] Ltd. Leighton Buzzard, England).
2. Filters can be treated with a UV-crosslinker instead of baking. If one is available, the time can be greatly reduced.
3. Oligonucleotide probes may need cleaning, to remove unincorporated nucleotides, before labeling. If the cleaning is not done by a commerical lab that may have synthesized the probe, you can use kits, like the MERmaid Kit from BIO 101, Inc.
4. The digoxygenin system of labeling and dectection was previously called the Genius system and was produced by Boerhinger Mannheim prior to their merger with Roche to form Roche Molecular Biochemicals.
5. An incubator with good stability and accuracy in the range of required temperatures is vital. Styles include: conventional shaking incubators and those expressly designed for hybridization. The latter type can be obtaining with reusable hybridization vessels which can be used in place of heat-sealable plastic bags. In either case, the temperature adjusts slowly over time. Wait 10–15 min after last adjustment to make sure temperature has stabilized. Temperature is critical for hybridization and depends on probe size and sequence.
6. When using heat sealable bags, make sure to remove as many air bubbles as pos-

sible. It is often necessary to partially seal the bag, squeeze out air bubbles through a small opening in the corner, and lastly seal that corner.

7. Kodak X-ray film can be used (such as Biomax or X-omat). The best results may be obtained using Lumi-film Chemiluminscent Detection Film (Roche Molecular Biochemicals) or equivalent.
8. Film can also be readily developed using an automated developer by following the manufacturer's directions.

References

1. Leff, L. G., Kernan, R. M., McArthur, J. V., and Shimkets, L. J. (1995) Identification of aquatic *Burkholderia (Pseudomonas) cepacia* by hybridization with species-specific rRNA gene probes. *Appl. Environ. Microbiol.* **61,** 1634–1636.
2. Braun-Howland, E. B., Vescio, P. A., and Nierzwicki-Bauer, S. A. (1993) Use of simplified cell blot technique and 16S rRNA-directed probes for identification of common environmental isolates. *Appl. Environ. Microbiol.* **59,** 3219–3224.
3. Schleifer, K. H., Amann, R., Ludwig, E., Rothemund, C., Springer, N., and Dorn, S. (1992) Nucleic acid probes for the identification and in situ detection of *Pseudomonas*. In *Pseudomonas: Molecular Biology and Biotechnology* (Galli, E., Silver, S., and Withold, B., eds.), American Society for Microbiology, Washington, DC,
4. Leff, L. G., Leff, A. A., and Lemke, M. J. (1998) Seasonal changes in planktonic bacterial assemblages of two Ohio streams. *Freshwater Biol.* **39,** 129–134.
5. Lemke, M. J., Brown, B. J., and Leff, L. G. (1997) The response of three bacterial populations to pollution in a stream. *Microb. Ecol.* **34,** 224–231.
6. Lemke, M. J. and Leff, L. G. (1999) Bacterial populations in an anthropogenically disturbed stream: comparison of different seasons. *Microb. Ecol.* **38,** 234–243.
7. Govan, J. R. W., Hughes, J. E., and Vandamme, P. (1996) *Burkholderia cepacia*: medical, taxonomic, and ecological issues. *J. Med. Microbiol.* **45,** 395–407.
8. Isles, A., Maclusky, I., Corey, M., Gold, R., Prober, C., Fleming, P., and Levison, H. (1984) *Pseudomonas cepacia* infection in cystic fibrosis: an emerging problem. *J. Pediatr.*(St. Louis) **104,** 206–210.
9. Vandamme, P., Mahenthiralingam, E., Holmes, B., Coenye, T., Hoste, B., De Vos, P., Henry, D. and Speert, D. P. (2000) Identification and population structure *of Burkholderia stabilis* sp. nov. (formerly *Burkholderia cepacia* Genomovar IV). *J. Clin. Microbiol.* **38,** 1042–1047.
10. MuKwaya, G. M. and Welch, D. F. (1989) Subgrouping of *Pseudomonas cepacia* by cellular fatty acid composition. *J. Clin. Microbiol.* **27,** 2646–2649.
11. Shimkets, L. J. and Asher, S. J. (1988) Use of recombination techniques to examine the structure of the *csg* locus of *Myxococcus xanthus*. *Mol. Gen. Genet.* **211,** 63–71.

18

In Situ Hybridization of *Burkholderia cepacia* Cells

Christopher J. McNamara and Laura G. Leff

1. Introduction

Burkholderia cepacia has attracted the attention of scientists from diverse fields owing to its unique characteristics and ubiquitous nature. This organism is an opportunistic human pathogen ***(1)***, one of the relatively few bacteria containing more than one chromosome ***(2)***, and able to utilize more compounds as a sole source of carbon and energy than any other known bacterium ***(3)***. Initially described as a plant pathogen ***(4)***, *B. cepacia* has been found in environments ranging from hospitals ***(5)*** to water supply systems ***(6)***, and is a common component of the bacterial assemblage in a wide range of natural habitats including freshwater ***(7–9)***, freshwater sediments (C. J. McNamara, *personal communication)*, soil ***(10)***, and leaves (C. J. McNamara, *personal communication*).

Accurate enumeration of bacteria (such as *B. cepacia*) in environmental samples is essential to the study of microbial ecology. Fluorescent nucleic acid stains, such as acridine orange (AO) and 4',6-diamidino-2-phenylindole (DAPI), afford an easy and rapid means to enumerate environmental bacteria ***(11–13)***, but provide no information about the taxonomic composition of samples. In addition, they have revealed that <10% (and frequently <1%) of environmental bacteria are culturable. Because traditional methods for the identification of bacteria rely on metabolic tests of pure cultures, the majority of environmental bacteria are unidentifiable using conventional methods.

The development of DNA-based techniques has provided new methods for the identification and quantification of environmental microorganisms ***(14,15)***. *In situ* hybridization, in particular, has allowed the quantification of nonculturable bacteria in environmental samples ***(15–20)***. Early methodologies describing *in situ* hybridization techniques used radiolabeled oligonucle-

From: *Methods in Molecular Biology, vol. 179: Gene Probes: Principles and Protocols*
Edited by: M. Aquino de Muro and R. Rapley © Humana Press Inc., Totowa, NJ

otides and visualized cells using microautoradiography ***(16)***. Because of the difficulties inherent in microautoradiography ***(14)***, fluorochromes were developed to label oligonucleotides, thereby allowing visualization of cells by epifluorescence microscopy. However, polycarbonate filters commonly used with general nucleic acid stains such as AO and DAPI autofluoresce when used in conjunction with fluorochromes for *in situ* hybridization ***(20)***. Alternative methods to concentrate samples prior to hybridization were developed, such as centrifugation ***(17)*** and filtration onto polycarbonate filters ***(21)***, each followed by transfer to glass slides. Unfortunately, both of these methods introduce significant opportunities for the loss of cells ***(20)***.

In the method described below for the *in situ* hybridization of *B. cepacia* cells, samples are concentrated onto Anodisc filters, which produce a low level of autofluorescence. All subsequent steps can be accomplished on the same filter, eliminating the need to transfer the sample and thereby reducing the potential loss of cells and increasing the accuracy with which environmental bacteria can be enumerated ***(20)***.

2. Materials

1. Incubator (*see* **Note 1**).
2. Vacuum filtration apparatus (25-mm diameter; includes glass funnel, support base, clamp, and sidearm flask) and vacuum pump.
3. Epifluorescence microscope with Texas Red filters (e.g., filter set 41004, Chroma Tech Corp., Brattleboro, VT; exciter HQ560/55, dichroic Q595LP, emitter HQ645/75).
4. 0.2-μm Filtered, autoclaved dH_2O (*see* **Note 2**).
5. 1× Phosphate-buffered saline (PBS): 7.6 g of NaCl, 1.9 g of Na_2HPO_4, 0.7 g of NaH_2PO_4 in 1 L, pH 7.2.
6. 8% Paraformaldehyde: Add 1.5 mL of 1 *M* NaOH to 800 mL of dH_2O, followed by 80 g of paraformaldehyde. In a fume hood, heat until dissolved and adjust volume to 1 L.
7. 0.1% Nonidet P-40 (Sigma Chemical, St. Louis, MO).
8. Hybridization buffer: 6× SSC (1× SSC: 0.15 *M* NaCl, 0.015 *M* sodium citrate, pH 7.0) 0.02 *M* Trizma base, pH 7.0) 0.1% sodium dodecyl sulfate (SDS); and 0.01% polyadenylic acid (*see* **Note 3**).
9. Wash buffer: 0.9 *M* NaCl; 0.02 *M* Trizma base, pH 7.2; 0.1% SDS.
10. Oligonucleotide labeled 5' with Texas Red. Sequence: 5'-CCTCTGTTCCGACCA-3' for the 16S rRNA of *B. cepacia* ***(22)*** (*see* **Note 4**).
11. Anodisc Filters, with polypropylene ring (25 mm diameter, 0.2 μm pore size, Whatman, Maidstone, UK).
12. Nitrocellulose filters 25 mm, 0.2 μm pore size (Millipore, Bedford, MA).
13. Petri dishes.
14. Zip-Lock bags (Dow Brands, Indianapolis, IN).

15. Kimwipes (Kimberly-Clark, Roswell, GA).
16. Filter paper (Fisher Brand Qualitative P8 Filter Paper, Fisher Scientific, Pittsburgh, PA).
17. 1.5-mL amber colored microcentrifuge tubes.
18. Type FF nonfluorescing immersion oil.
19. Glass slides and coverslips.

3. Methods

3.1. Sample Preservation and Probe Dilution

1. Prepare preservative by mixing equal volumes 1× PBS and 8% paraformaldehyde. Preserve samples with ice cold 0.5× PBS, 4% paraformaldehyde (preservative volume to sample volume ratio = 3:1). Store samples at 4°C until hybridization ***(18).***
2. Lyophilized probes should be stored at –20°C after being received from the manufacturer. Hydrate lyophilized probe using dH_2O to 100 ng/μL (*see* **Note 5**).
3. Prior to hybridization, dilute the probe to 50 ng/μL with dH_2O, and then to 5 ng/μL with hybridization buffer *(21)* (*see* **Note 6**).

3.2. Preparation of Hybridization Chambers

1. Set incubator temperature to 49°C (*see* **Note 7**).
2. Place diluted probe (5 ng/μL) and wash buffer in incubator.
3. Prepare chambers to prevent probe from evaporating during hybridization. Role a Kimwipe into a small ball, place in a ziplock bag, saturate with wash buffer, seal shut, and place in the incubator. Cut a piece of filter paper to fit the lid of a Petri dish and wet with about 1.0 of mL wash buffer so that it will stick to the lid of the Petri dish. Retain the Petri dish for the following steps.

3.3. Sample Concentration and Hybridization

1. Place a nitrocellulose filter on the filtration apparatus support base and wet with dH_2O (*see* **Note 8**).
2. Place a 0.2 μm pore size Anodisc filter on top of the nitrocellulose filter and clamp the funnel over both filters.
3. Concentrate sample onto filter by aspirating at 15 kPa vacuum (5 mm Hg).
4. With vacuum pump still running, rinse sample with 1.0 mL of dH_2O followed by 1.0 mL of 0.1% Nonidet P-40.
5. Place Anodisc on the bottom of the Petri dish.
6. Pipet 40 μL of probe (5.0 ng/μL) onto the top of each filter.
7. Place Petri dish inside a ziplock bag with Kimwipe and incubate in the dark for 4 h at 49°C.

3.4. Washing and Mounting Filters

1. At the end of the hybridization period, remove from incubator and place filters on filtration apparatus support base (*see* **Note 9**).

2. Rinse each filter twice with 400 µL of wash buffer.
3. Place in a new Petri dish and pipet 80 µL of wash buffer onto the top of the filter (*see* **Note 10**).
4. Return to the incubator for 10 min.
5. Repeat **steps 2–4** of this subheading.
6. Rinse each filter twice with 400 µL of dH_2O.
7. Place a drop of immersion oil on a glass slide and place filter in the oil with the sample side up. Add a second drop of oil on top of the filter followed by cover glass.
8. Enumerate cells using epifluorescence microscopy.

4. Notes

1. Temperature readouts can vary significantly among incubators. Because slight differences in temperature can significantly effect the outcome of the hybridization, place a good quality, total immersion mercury thermometer inside the incubator.
2. All solutions must be filtered to remove particles in the same size range as bacteria and autoclaved to prevent bacterial growth.
3. When making the hybridization and the wash buffers, use stock solutions of 20× SSC, 3 *M* NaCl, 1 *M* Trizma, and 10% SDS. A precipitate will form when preparing the buffers. Heat buffers just until the precipitate dissolves prior to adjusting the pH to the desired level (pH 7.0 for hybridization buffer, pH 7.2 for wash buffer).
4. When purchasing the oligonucleotide, specify polyacrylamide gel electrophoresis (PAGE) purification and lyophilization prior to shipment. PAGE purification will reduce the yield of probe below that of other purification methods (e.g., HPLC), but is necessary to ensure the quality of the probe and the subsequent reliability of the hybridization.
5. In general, oligonucleotides should be rehydrated using TE (pH 7.5–8.0) in order to prevent autocatalytic degradation due to acidic conditions. However, the presence of EDTA can interfere with hybridizations. Therefore, rehydrate the probe in dH_2O and then immediately dilute and aliquot as described in **Note 6**. The hybridization buffer will prevent autocatalytic degradation due to low pH.
6. Because light will degrade the fluorochrome over time, the probe should be kept in the dark. In addition, repeated freezing and thawing will also degrade the probe. Dilute the probe to 5 ng/µL as described earlier and dispense 1.0-mL aliquots into 1.5 mL amber colored microcentrifuge tubes. Aliquots should then be stored at –70°C until use.
7. Stringency of the hybridization can be adjusted using slight changes in the temperature. We recommend running a blank as well as positive and negative controls with each hybridization. If the probe appears to have hybridized to the negative controls, increase the temperature (in 1°C increments). Likewise, if the positive control appears faint, decrease the temperature (again, in 1°C increments). Cultures to be used for controls can be purchased (e.g., from the Ameri-

can Type Culture Collection), grown to log phase in the appropriate medium, and preserved as described above. Cultures should be preserved while in log phase growth to ensure that most cells have a high number of ribosomes to which the probe can hybridize. However, it will then be necessary to dilute the culture before filtration and hybridization (a 1:100 dilution of the preserved culture usually works well).

8. The nitrocellulose filters are used to support the thin, fragile anodiscs. Filters with other pore sizes (e.g., 0.45 µm) are frequently used.
9. Owing to the polypropylene ring around the perimeter of the Anodisc, it is not necessary to attach the filter tower for the wash steps.
10. Petri dish lids with the filter paper can be re-used. However, the bottoms should be changed after each step to avoid returning the filters to an area with a high concentration of probe.

References

1. Isles, A., Maclusky, I., Corey, M., Gold, R., Prober, C., Fleming, P., and Levison, H. (1984) *Pseudomonas cepacia* infection in cystic fibrosis: an emerging problem. *J. Pediatr.*(St. Louis) **104,** 206–210.
2. Cheng, H. P. and Lessie, T. G. (1994) Multiple replicons constituting the genome of *Pseudomonas cepacia* 17616. *J. Bacteriol.* **176,** 4034–4042.
3. Lessie, T. G. and Gaffney, T. (1986) Catabolic potential of *Pseudomonas cepacia.* In *The Bacteria: A Treatise on Structure and Function*, Vol. X: *The Biology of Pseudomonas* (Sokatch, J. R., ed.), Academic Press, New York, pp. 439–481.
4. Palleroni, N. J. (1992) Human- and animal-pathogenic pseudomonads. In *The Prokaryotes*, Vol. III (Balows, A., Truper, H.G., Dworkin, M., Harder, W., and Schleifer, K.H., eds.) Springer-Verlag, New York, pp. 3086–3103.
5. Randall, C. (1980) The problem of *Pseudomonas cepacia* in a hospital. *Can. J. Public Health* **71,** 119–123.
6. Koenig, D. W. and Pierson, D.L. (1997) Microbiology of the space shuttle water system. *Wat. Sci. Technol.* **35,** 59–64.
7. Leff, L. G., Kernan, R. M., McArthur, J. V., and Shimkets, L. J. (1995) Identification of aquatic *Burkholderia (Pseudomonas) cepacia* by hybridization with species-specific rRNA gene probes. *Appl. Environ. Microbiol.* **61,** 1634–1636.
8. Lemke, M. J., Brown, B. J., and Leff, L. G. (1997) The response of three bacterial populations to pollution in a stream. *Microb. Ecol.* **34,** 224–231.
9. Lemke, M. J. and Leff, L. G. (1999) Bacterial populations in an anthropogenically disturbed stream: comparison of different seasons. *Microb. Ecol.* **38,** 234–243.
10. Bevivino, A., Sarroccom, S., Dalmastri, C., Tabacchioni, S., Cantale, C., and Chiarini, L. (1998) Characterization of a free-living maize-rhizosphere population of *Burkholderia cepacia*: effect of seed treatment on disease suppression and growth promotion of maize. *FEMS Microbiol. Ecol.* **27,** 225–237.
11. Daley, R. J. (1979) Direct epifluorescence enumeration of native aquatic bacteria: uses, limitations, and comparative accuracy. In *Native Aquatic Bacteria: Enumeration, Activity, and Ecology, ASTM STP 695* (Costerton, J. W. and Colwell, R. R., eds.), American Society for Testing and Materials, pp. 29–45.

12. Newell, S. Y., Fallon, R. D., and Tabor, P. S. (1986) Direct microscopy of natural assemblages. In *Bacteria in Nature,* Vol. II. *Methods and Special Applications in Bacterial Ecology* (Poindexter, J. S. and Leadbetter, E. R., eds.), Plenum Press, New York, pp. 1–48.
13. Kepner, R. L. and Pratt, J. R. (1994) Use of fluorochromes for direct enumeration of total bacteria in environmental samples: past and present. *Microbiol. Rev.* **58,** 603–615.
14. Saylor, G. S. and Layton, A. C. (1990) Environmental application of nucleic acid hybridization. *Ann. Rev. Microbiol.* **44,** 625–648.
15. Head, I. M., Saunders, J. R., and Pickup, R. W. (1998) Microbial evolution, diversity, and ecology: a decade of ribosomal RNA analysis of uncultivated microorganisms. *Microb. Ecol.* **35,** 1–21.
16. Giovannoni, S. J., DeLong, E. F., Olsen, G. J., and Pace, N. R. (1988) Phylogenetic group-specific oligodeoxynucleotide probes for identification of single microbial cells. *J. Bacteriol.* **170,** 720–726.
17. DeLong, E. F., Wickham, G. S., and Pace, N. R. (1989) Phylogenetic stains: ribosomal RNA-based probes for the identification of single cells. *Science* **243,** 1360–1363.
18. DeLong, E. F (1993) Single-cell identification using fluorescently labeled, ribosomal RNA-specific probes. In *Handbook of Methods in Aquatic Microbial Ecology* (Kemp, P. F., Sherr, B. F., Sherr, E. B., and Cole, J. J., eds.) Lewis, Boca Raton, FL, pp. 285–294.
19. Hahn, D., Amann, R. I., and Zeyer, J. (1993) Whole-cell hybridization of *Frankia* strains with fluorescence- or digoxigenin-labeled, 16S rRNA targeted oligonucleotide probes. *Appl. Environ. Microbiol.* **59,** 1709–1716.
20. Lemke, M. J., McNamara, C. J., and Leff, L. G. (1997) Comparison of methods for the concentration of bacterioplankton for *in situ* hybridization. *J. Microbiol. Methods* **29,** 23–29.
21. Hicks, R. E., Amann, R. I., and Stahl, D. A. (1992) Dual staining of natural bacterioplankton with 4',6-diamidino-2-phenylindole and fluorescent oligonucleotide probes targeting kingdom-level 16S rRNA sequences. *Appl. Environ. Microbiol.* **58,** 2158–2163.
22. Braun-Howland, E. B., Vescio, P. A., and Nierzwicki-Bauer, S. A. (1993) Use of simplified cell blot technique and 16S rRNA-directed probes for identification of common environmental isolates. *Appl. Environ. Microbiol.* **59,** 3219–3224.

19

Detection of Polychlorinated Biphenyl-Degrading Organisms in Soil

John Sanseverino, Alice C. Layton, and Gary S. Sayler

1. Introduction

Polychlorinated biphenyls (PCBs) are a group of 209 congeners consisting of a biphenyl ring with 1–10 chlorines. In the United States, PCBs, manufactured under the trade name Aroclor, are ubiquitous and recalcitrant pollutants in the environment. PCBs have been shown to biomagnify in the food chain and are associated with chronic health effects *(1,2)*.

Biological degradation of PCBs was first reported in 1973 with the work of Ahmed and Focht *(3)*. Although no single organism has been isolated which completely degrades all possible PCB congeners, evidence exists indicating PCBs may be biologically degraded in the environment by complementary anaerobic and aerobic processes *(4)*. In anaerobic biodegradation, PCBs are dechlorinated through reductive dehalogenation which serves as an energy source for the microorganism. This process favors more highly chlorinated PCB congeners and results in less chlorinated PCBs *(5)*. Aerobic biodegradation favors the less chlorinated PCB congeners and results in ring cleavage of the biphenyl backbone.

The population density and activity of contaminant-degrading microorganisms are among the key diagnostic parameters used in predicting field rate of contaminant biodegradation *(6,7)*. The measurement of contaminant-degrading microorganisms using traditional cultivation-based laboratory techniques often underestimates heterotrophic and specific bacterial populations by one to two orders of magnitude *(8)*. DNA:DNA hybridization between specific gene probes and DNA extracted from soil bacterial populations allows for more accurate identification and quantification of specific bacterial populations because relevant catabolic genes can be targeted.

From: *Methods in Molecular Biology, vol. 179: Gene Probes: Principles and Protocols*
Edited by: M. Aquino de Muro and R. Rapley © Humana Press Inc., Totowa, NJ

1.1. Selection of Gene Probes for PCB Degrading Organisms

The application of molecular methods for monitoring biodegradation is dependent on the characterization of the relevant catabolic genes. Currently, catabolic genes available for monitoring PCB biodegradation belong to the biphenyl operon. The biphenyl operon encodes the genes for 4 enzymes mediating the conversion of biphenyl and chlorinated biphenyls to benzoic acid and chlorobenzoic acids ***(9)***, and often an additional three enzymes for the conversion of 2-hydroxypenta-2,4-dienoate to pyruvate and acetyl-CoA and pyruvate, and glutathione-*S*-transferase ***(10,11)***. The well-studied biphenyl operon was originally cloned and sequenced from the Gram-negative bacterium *Pseudomonas* LB400 (now classified as *Burkholderia cepacia* LB400 ***(12,13)*** and *Pseudomonas pseudoalcaligenes* KF707 ***(14,15)***. Since then, the biphenyl operon from a number of Gram-negative species including *P. putida* OU83 ***(16–18)***, *P. testosteroni* B-356 ***(19–21)***, *Ralstonia eutrophus* ENV307 (formerly *Pseudomonas* sp. ENV307; ***22***) and Gram-positive *Rhodococcus* species ***(23–28)*** have been cloned and sequenced. Another recently identified, but not cloned operon is from the *Bacillus* group ***(29)***. The biphenyl operon is conserved between certain groups of bacteria and is related to other operons that were originally isolated in conjunction with the ability to degrade toluene and isopropyl benzene (**Fig. 1**). Biphenyl operons also vary in their substrate range with regard to PCB congeners ***(30)***. Some biphenyl dioxygensases are capable of degrading PCB congeners with only one or two chlorines, whereas some biphenyl dioxygenases such as from *Burkholderia* LB400 ***(31)***, and the *Rhodococcus* strains ***(32)*** can initiate degradation of congeners with up to six chlorines. Other lignin degrading microorganisms such as *Phanerochaete chrysosporium* and *Sphingomonas paucimobilis* strain SYK-6 may also degrade PCBs ***(33,34)***.

Gene probes and PCR primers for the detection of PCB-degrading bacteria have been made to target the *bphC* gene encoding the enzyme 2,3-dihydroxybiphenyl dioxygenase from Gram-negative PCB-degrading bacteria ***(17,35–38)***, *bphA1* gene encoding the large subunit of the iron-sulfur protein component of the biphenyl dioxygenase ***(36)*** and *bphK* ***(10)***. The problem associated with using the *bphC* gene as a probe is that there are other dioxygenases (such as naphthalene and toluene) present in decomposing bacteria not capable of PCB degradation that are similar enough to the *bphC* gene to cross-hybridize ***(37)***. The large subunit of the iron-sulfur protein component of the biphenyl dioxygenase gene (*bphA1*) determines PCB congener specificity and thus may make a better gene probe than the *bphC* gene for distinguishing PCB-degrading organisms ***(31,39)***.

Although all of the sequenced *bphA1* genes are at least 70% similar at the DNA level, at least five gene probes (one for each PCB group) would be needed

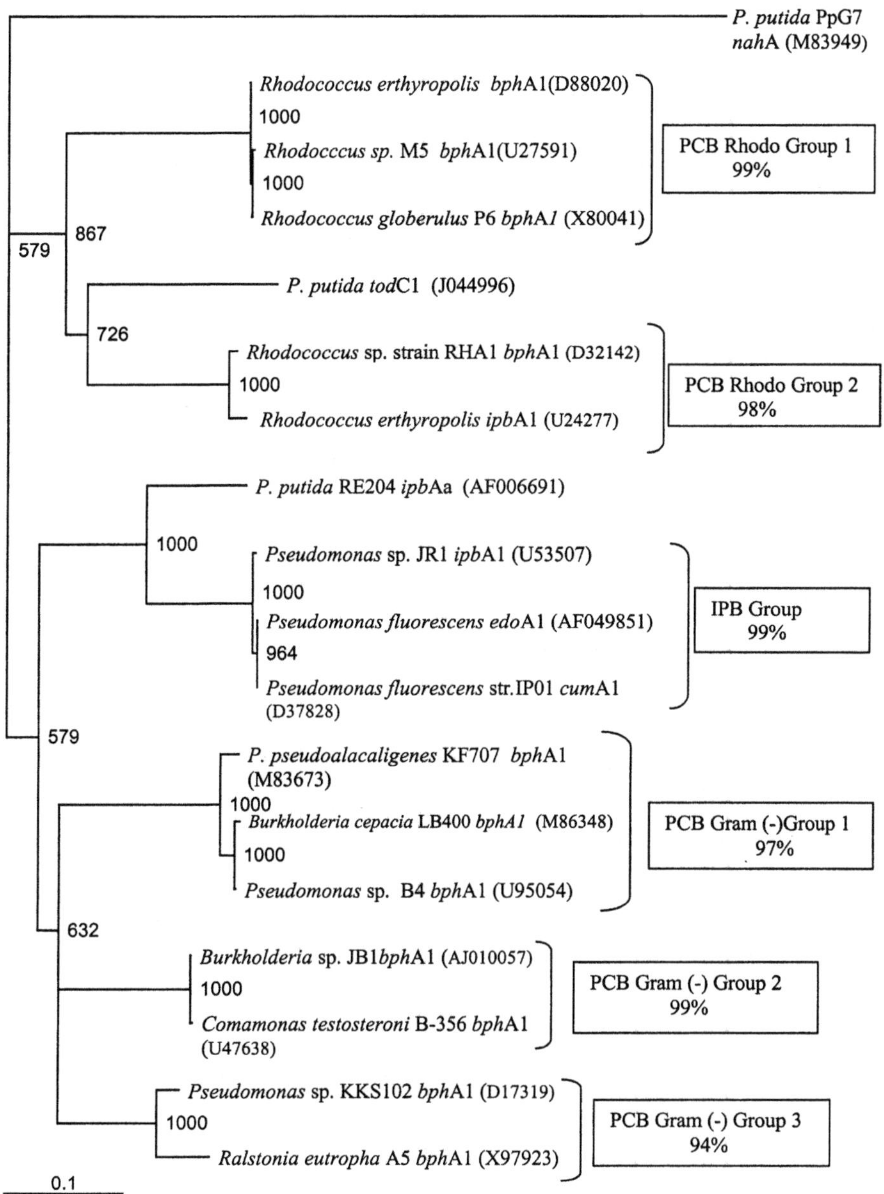

Fig. 1. Relationships of DNA sequences from the large subunit of the iron sulfur protein component of the biphenyl dioxygenase gene (*bphA1*) and related dioxygenases. Eight hundred and fifty basepairs were aligned using Clustal W and trees were constructed using the neighbor joining method and 1000 bootstrap analysis were performed ***(42)***. Trees were viewed in the TREEVIEW program ***(43)*** with the *nahA* gene from *P. putida* PpG7 used as an outgroup. Genbank accession numbers are provided in parentheses and the percent similarity between each group is provided in parentheses.

Table 1
Primers for the Amplification of *bphA1* Genes from the Biphenyl Operons of PCB-Degrading Bacteria

Target group[a]	Forward primer Reverse primer [b]	Melting temperature (°C)
PCB Gram-negative Group 1	F-5'GTGAAGTGGGTTACCAATTGGA3' R-5'GGCGATATTCTTCCTTGATCTC3'	64 64
PCB Gram-negative Group 2	F-5'TTAGGTGGTCCCGCAACTGGA3' R-5'AGCGGAATTCCTCCTTGATATC3'	66 64
PCB Gram-negative Group 3	F-5'TGACGTTCAAGCGTCGCTGGA3' R-5' GCGGAACTCTTCCTTGATGTC 3'	66 64
PCB Group 1	F-5'TCAATTGGGTCGCACCTCAAC3' R-5'CCTGTACTCCTCCTTGATCTC3'	64 64
PCB Group 2	F-5'TGACTGACGTGCAATGTGAACC3' R-5'GGCGGTATTCCTCTTTCATCTC3'	66 66

[a] Organisms in each target group are found in **Fig. 1**.

[b] Primers were designed by aligning available *bphA1* sequences in Clustal W. All forward primers were designed from the region of *bphA1* 5–40 bp downstream from the initial start codon. All reverse primers were designed from the sequence region approx 1100 bp from the *bphA1* start codon and are the reverse compliment of the aligned sequences. Primers were designed to anneal at approximately the same temperature. Primer sequences were checked using Blast analysis (NCBI). The product size for each primer pair is approx 1050 bp.

to identify all of the PCB-degrading organisms in the environment with available DNA sequences and to distinguish between PCB degraders and other aromatic compound degraders such as *P. pudita* F1 (tod) and the IPB Group (**Fig. 1**). Multiple probes are needed because hybridization temperatures need to be lowered 0.5–1.4°C for each 1% mismatch ***(40)***, so that for sequences with less than 90% similarity the necessary hybridization temperature may result in high background. In addition to molecular analysis, chemical analysis such as PCB growing cell or resting cell assays are needed to determine the substrate specificity range of the bacteria in a sample ***(30,36,41)***, because the *bphA1* gene from narrow PCB congener substrate range and broad PCB congener substrate range degraders are intermixed (e.g., in Gram-negative Group 1, LB400 is a broad substrate range degrader whereas KF707 a narrow substrate range PCB degrader).

A complete set of gene probes needed to detect currently isolated and sequenced biphenyl operons have not been published. Therefore, potential primers for the amplification of the *bphA1* gene corresponding to the PCB operon groups identified in **Fig. 1** are provided in **Table 1**. Other primers for the amplification of PCB-degrading genes can be designed by aligning DNA sequences available in Genbank (NCBI: www.ncbi.nlm.nih.gov) or EMBL using alignment programs such as Clustal W. Primers can be designed by visual comparison of aligned sequences or using programs such as Primer 3 (www.genome. wi.mit.edu).

1.2. Objectives

The objective of this chapter is to familiarize the reader with procedures for analyzing an environmental sample for *bphA* genes. As discussed previously, to capture all *bph* operons, a suite of *bph* probes representing the different *bph* operons will be needed. The methods in this chapter include DNA extraction from soil, probe generation, hybridization, and data analysis. These are general methods that may be used for analysis of any specific gene of interest. The user may have to modify parameters, such as hybridization temperatures, polymerase chain reaction (PCR) mixture, and reaction conditions, for different probes.

2. Materials

2.1. DNA Extraction and Purification (Subheading 3.1.)

1. Bead-beater (Bio Spec, Bartlesville, OK).
2. 0.1-mm Glass beads.
3. Rotary water bath.
4. 250-mL Centrifuge bottles.
5. Dialysis tubing (mol wt cutoff 6000–8000).
6. 0.12 M Na_2HPO_4, pH 8.0.
7. 5% (w/v) Sodium dodecyl sulfate (SDS).
8. 2 *M* Na acetate, pH 5.2.
9. Isopropanol.
10. TE buffer: 10 m*M* Tris-HCl, 1 m*M* EDTA, pH 7.5.
11. Tris-saturated phenol, pH 8.0.
12. Chloroform–isoamyl alcohol (24:1).
13. Absolute ethanol.

2.2. Polymerase Chain Reaction Amplification and Probe Preparation (Subheading 3.2.–3.5.)

1. Polymerase Chain Reaction amplification kit.
2. PCR tubes.
3. TA Cloning Kit (Invitrogen, San Diego, CA).
4. [α-^{32}P]dCTP (ICN, Costa Mesa, CA).
5. Thermal cycler.
6. Nuc-Trap Columns (Stratagene, La Jolla, CA).

7. STE buffer: 10 m*M* Tris-HCl, 100 m*M* NaCl, 1 m*M* EDTA, pH 8.0.
8. *Hin*dIII restriction enzyme.
9. 0.8% Agarose gel in TBE buffer.
10. 1.0% Agarose gel in TBE buffer.

2.3. Slot Blot Hybridization (Subheading 3.6.)

1. Biotrans Nylon membrane (ICN, Costa Mesa, CA).
2. Slot Blot apparatus.
3. Whatman 3 MM Filter paper.
4. 2× SSC: 17.5 g of NaCl, 8.82 g of Na citrate, pH 7.0, adjust to 1 L with distilled water.
5. 0.5 *M* NaOH (made fresh).

2.4. Hybridization (Subheading 3.7.)

1. Heat-sealable pouches and sealer.
2. Rotary water bath (55°C).
3. Phospho Imager or X-ray film and developer.
4. Hybridization solution: 0.5 *M* NaH_2PO_4, 1 m*M* EDTA, pH 8.0, 7% w/v SDS, 1 L of dH_2O; pH 7.2 ***(44)***.
5. High stringency wash buffer: 10 m*M* NaCl, 20 m*M* Tris-HCl, 1 m*M* EDTA, 0.5% SDS pH 7.0–8.0.
6. Whatman 3 MM filter paper.

3. Methods

3.1. Extraction and Purification of DNA from Soil (see Note 1)

Various protocols for extracting and purifying DNA from soil have been developed ***(45–49)***. The method described here is based on Ogram et al. ***(45)*** as modified by Stapleton et al. ***(50)*** and Stapleton ***(51)***. This method yields high quality DNA suitable for PCR amplification.

1. Weigh 50 g of soil and transfer to a bead beater blender. Add 125 mL of 0.12 *M* Na_2HPO_4 and 25 mL of 5% SDS.
2. Seal and incubate at 70°C with gentle shaking for 1 h. A rotary water bath set at 50 rpm is suitable. Periodically, invert the blender for top-to-bottom mixing.
3. Add 25 g of 0.1-mm diameter glass beads.
4. Blend twice for 2.5 min with a 0.5-min rest in between.
5. Let the mixture cool on ice for 2 min.
6. Add 100 ng of λ standard (*see* **Subheading 3.3.**).
7. Transfer the mixture to a 250 mL centrifuge bottle. Centrifuge at 5500*g* for 25 min at 10°C. Decant and save the supernatant (Supernatant 1) in a clean 250-mL centrifuge bottle.

8. Add 125 mL of 0.12 *M* Na_2HPO_4 to the pellet and incubate at 70°C with gentle shaking for 20 min.
9. Centrifuge at 5500*g* for 25 min at 10°C. Decant and save the supernatant in a clean 250 mL centrifuge bottle (supernatant 2).
10. Repeat **steps 7** and **8**. If additional particulate material is present, centrifuge the supernatants at 11,500*g* for 30 min.
11. Precipitate DNA by adding 0.1 volume of 2 *M* Na acetate, pH 5.2, and an equal volume isopropanol to each supernatant. Incubate overnight at –20°C.
12. Pellet the DNA by centrifugation at 11,500*g* for 30 min at 4°C. Discard the supernatants.
13. Dry the pellets under vacuum. This is accomplished by removing the cap from the bottle and sealing with parafilm. Place pinholes in the parafilm with a toothpick. Place the sealed centrifuge bottles in a SpeedVac and turn on the vacuum only.
14 Resuspend the pellets in 20 mL of TE buffer.
15. Pool the resuspended pellets in a dialysis bag and dialyze overnight at 4°C against TE buffer to remove excessive salt.
16. Extract the dialysate with an equal volume of Tris-saturated phenol, pH 8.0. Remove the aqueous phase and repeat.
17. Extract the aqueous phase twice with chloroform–isoamyl alcohol. Remove the aqueous phase to a clean centrifuge bottle.
18. Precipitate DNA by addition of 0.1 volume of 2 *M* Na acetate and 2 volumes of absolute ethanol at –20°C overnight.
19. Pellet the DNA by centrifugation at 11,500*g* for 30 min at 4°C. Discard the supernatants and dry the DNA as described in **step 13**.
20. Resuspend pellets in 1 mL of sterile TE buffer and determine the DNA concentration. Store at –20°C until used.

3.2. Preparation of bphA Standards for Gene Probing (see Note 2)

1. Set-up the following reaction mixture for amplification of double-stranded *bphA1*:
   Overlay the reaction with mineral oil or paraffin wax beads. The DNA template is a chromosomal preparation of *Burkolderia cepacia* LB400.

Component	Volume
10X PCR buffer	10 µL
50 m*M* $MgCl_2$	3 µL
DNA template	1 µg
10 m*M* dNTP mix	2.5 µL
bphA Primer 1 (5 ng/µL)	2.5 µL
*bph*A Primer 2 (5 ng/µL)	2.5 µL
Taq polymerase	1 µL
Sterile distilled water	to 100 µL

2. The following thermal cycler guidelines may be used for amplifying the probe:

Step	Temperature (°C)	Time
Denaturation	94	30 s
Annealing	60	20 s
Elongation	72	2 min
		Repeat for 35 cycles
Extended elongation	72	10 min
Soak	4	—

3. Run the amplification product on a 0.8% agarose gel in TBE buffer to confirm the presence of the ~1.0-kb band.
4. Repeat the amplification using 1 µL of the ~1.0-kb amplification product as the DNA template.
5. Repeat the amplification a third time using the amplification product from **step 4**.
6. Quantify the amplification product by UV detection or fluorescence detection.
7. Prepare 2-mL working stocks of the *bphA1* fragment in the following concentrations (per 100 µL): 0.01, 0.03, 0.1, 0.3, 1, 3, 10, 30, and 100 ng.
8. Store at –20°C until ready for use.

3.3. Preparation of λ Internal Standards

1. Amplify a λ 500-bp fragment using control primers and template from a Perkin Elmer PCR reagent kit according to manufacturer's protocols.
2. Run on a 1.0% gel to confirm proper band size.
3. Clone into the TA Cloning Vector (Invitrogen, San Diego, CA) according to the manufacturer's protocols.
4. Choose a positive clone and perform a large-scale plasmid preparation using a procedure such as Promega ***(52)***.
5. Prepare a working solution (100 ng of λ fragment/100 mL) of plasmid DNA.
6. Prior to use as an internal standard in **Subheading 3.1., step 6**, linearize the plasmid with *Hin*dIII to produce a single DNA fragment (0.5-kb λ + 3.9-kb vector) (*see* **Note 3**).

3.4. Preparation of λ Standards for Gene Probing

The protocols used in **Subheading 3.2.** should be used to prepare a series of 500-bp fragment λ standard for use in slot blot hybridization.

3.5. Amplification and Purification of ^{32}P-Labeled Single-Stranded bphC Probes (See Note 4)

The following procedure assumes the user is authorized to handle radioisotopes and understands the hazards and takes the necessary precautions.

1. Set up the following reaction mixture for amplification of single-stranded probe:

Component	Volume
10× PCR buffer	10 µL
50 m*M* $MgCl_2$	3 µL
Double-stranded DNA template	1 µg
10 m*M* dATP	2 µL
10 m*M* dTTP	2 µL
10 m*M* dGTP	2 µL
[α-^{32}P]dCTP	10 µL
Primer (5 ng/µL)	2.5 µL
Taq polymerase	1 µL
Sterile distilled water	to 100 µL

   Overlay the reaction with mineral or paraffin wax beads. The DNA template is double-stranded *bphA* generated in **Subheading 3.2.** A second reaction should also be set-up using the double-stranded λ fragment as the DNA template.
2. The following thermal cycler guidelines may be used for amplifying the probe:

Step	Temperature (°C)	Time
Denaturation	94	30 s
Annealing	60	20 s
Elongation	72	2 min
	Repeat for 35 cycles	
Extended elongation	72	10 min
Soak	4	—

3. Equilibrate a push column (NucTrap, Stratagene Corporation) with 70 µL of STE buffer and push.
4. Add 50 µL of PCR extract to the column and push. Be careful to avoid the mineral oil overlay if present.
5. Repeat with the remaining PCR extract.
6. Elute the labeled single-stranded probe with 70 µL of STE buffer and collect in a 1-mL tube. Discard column.
7. Bring volume up to 1 mL with TE buffer.
8. Count 5 µL in a scintillation counter. The total incorporated radioactivity in the 1-mL probe should not be less than 1,000,000 cpm.
9. The probe should be used immediately while the specific activity is high. If not, store at –20°C until ready to use. Probes that are stored for longer than 7 d should not be used and disposed of properly.

3.6. Slot Blot Hybridization (see Note 5)

1. Make several dilutions of the extracted DNA. The concentration of the target DNA should fall in the linear range of the DNA standards.
2. Heat 100 µL of extracted DNA + 400 µL of 0.5 *M* NaOH at 100°C for 10 min.
3. Heat 100 µL of the DNA standards + 400 µL of 0.5 *M* NaOH at 100°C for 10 min. Make replicate membranes with the appropriate standards and extracted DNA. One membrane should be loaded with the double-stranded *bph* standards prepared in **Subheading 3.2.** while a second membrane should be loaded with the λ standards.
4. Wet a sheet of membrane (Biotrans™ Nylon Membrane) and two filter papers for slot blot by immersing in distilled water. Label the membrane with a pencil before immersing in water.
5. Assemble the slot blot apparatus with the pre-wetted membrane putting the membrane at the top of the two filter papers. Rinse wells with 0.5 mL of TE buffer. Apply vacuum until wells are empty but not dry.
6. Apply the 0.5 mL of DNA sample to each appropriate well without vacuum.
7. Start vacuum until the wells are almost dry.
8. Rinse all wells by placing 0.5 mL of 0.4 *M* NaOH in each, then apply vacuum until all wells are dry.
9. Disconnect the vacuum, disassemble the apparatus, and rinse the membrane briefly in 2× SSC. Air-dry the membrane.
10. Bake the membrane at 80°C for 1 h. The membrane can be stored indefinitely at room temperature between two pieces of filter paper in a plastic ziploc bag.

3.7. Probing and Detection of DNA:DNA Hybrids (see Notes 6 and 7)

1. Place the membrane in a heat-sealable pouch. For a pint-size pouch add 10 mL of hybridization buffer. Remove air bubbles and seal.
2. Prehybridize the blots in a 55°C shaking (gentle) water bath for 1 h.
3. Open the bag by making a small cut in one corner. Add the labeled probe and reseal the bag with the heat sealer. Use at least 1×10^6 dpm of probe.
4. Return the pouch to the 55°C shaking (gentle) water bath for a minimum of 8 h; usually overnight incubation is adequate.
5. Remove the membrane from the pouch and place in a shallow plastic container. Add 500 mL of high-stringency wash buffer. Preheat the buffer to the hybridization temperature.
6. Return the plastic container to the 55°C shaking (gentle) water bath. Let incubate for 15 min.
7. Repeat this washing step three more times. Dispose of the high stringency wash buffer according to the facilities radiation disposal plan.
8. Remove excess moisture from the membrane by placing on Whatman 3MM filter paper (or something similar).
9. Visual and quantify the ^{32}P-labeled hybrids by autofluorography or by phosphor imaging.

3.8. Data Analysis

Quantitation of biphenyl-degrading organisms can be calculated from the slot blot hybridization data (*see* **Note 8**). Three calculations can be performed: DNA extraction efficiency, determination of total bacterial cell numbers, and normalization of the data. Applegate et al. ***(53)*** gives a discussion of these calculations.

3.8.1. Extraction Efficiency

The lysis efficiency is assumed to be 100%. Bead beating and high temperature SDS lysis is vigorous enough to warrant this assumption ***(54)***. Ogram et al. ***(45)*** determined by direct count that <90% of the cells were lysed by this procedure.

The λ fragment is used as the internal control to monitor recovery of DNA. λ was chosen because it was not found generally in the environment ***(53)***. If λ is present in the user's sample, then another λ fragment can be used to determine background values or the user can choose another appropriate internal control. Extraction efficiency is determined by the following equation:

$$(\text{ng of } \lambda \text{ recovered}/100 \text{ ng of } \lambda \text{ added}) = \text{extraction efficiency}$$

3.8.2. Estimation of Target Cell Population

An estimate of the number of biphenyl-degrading microorganisms that hybridized with the probe may be made using the following equation:

$$\begin{gathered}(\text{ng of hybridized DNA/ng of one target DNA molecule})\\(1/\text{number of target molecules})\ (1/\text{extraction efficiency})\\(1/\text{fraction of sample loaded onto blot})\\= \text{number of biphenyl-degrading cells}\end{gathered}$$

The nanograms of hybridized DNA is determined from the imaging software used to visualize and quantify the ^{32}P signal. The ng of the target DNA molecule is determined from the following equation:

$$\begin{gathered}(\text{number of base pairs in target})(660 \text{ g/mol of basepairs})\\(1 \text{ mole of base pairs}/6.02 \times 10^{23} \text{ molecules})(10^{9} \text{ ng/g})\\= \text{ng of one target DNA molecule}\end{gathered}$$

3.8.3. Percent Community Determination

To compare nucleic acid hybridization data from different samples, normalization of the biphenyl-degrading population to the total bacterial population can be performed. The total bacterial population is determined by hybridizing the extracted DNA with a universal 16S rDNA probe ***(55)***. The percent community is determined by the following equation ***(50)***:

$$\frac{(\text{ng of } \textit{bphA}/\text{copy number of } \textit{bphA})}{(\text{ng of 16S rDNA genes/copy number of 16S rDNA genes})} \times 100 = \text{percent community}$$

The copy number of 16S rDNA genes is 7 in *E. coli* ***(56)***. This number may vary depending on the genus/species ***(57)***.

4. Notes

1. The DNA extraction method is labor intensive, and other methods are available. However, this method provides large quantities of high quality DNA suitable for PCR, hybridization, and cloning. The DNA can be further purified by cesium chloride-ethidium bromide (CsCl–EtBr) ultracentrifugation if necessary. This may be necessary if the soil/sediment sample has a high organic matter. Humic materials copurify with DNA; however, low levels will not interfere with DNA hybridization. In addition, aromatic pollutants may also copurify with the DNA necessitating further purification with CsCl–EtBr ultracentrifugation.
2. If multiple bands are present in the amplification of the double-stranded DNA, then the PCR reaction may have to be optimized. Perform the amplification using Invitrogen's PCR Optimizer kit. In addition, increasing the annealing temperature of the annealing reaction may increase primer specificity and reduce mispriming and primer dimer formation. If multiple bands are still present, then gel purification may be necessary.
3. Linearized λ + vector standards are added to the soil extraction mixture instead of purified λ fragment. This λ fragment is 500 bp and may not purify with the bulk chromosomal DNA, although this has never been proven conclusively. The TA cloning vector + the λ fragment is 4.4 kb, which will copurify with the extracted DNA. In addition, if the vector and λ is linearized it will migrate with the extracted linear chromosomal DNA and not with any supercoiled plasmids that may also be in the mix.
4. Double-stranded probe templates may be cloned into the TA cloning vector for archiving. This user prefers to generate single-stranded probe from double-stranded template that is free of vector. This reduces the risk of generating false positives by labeling contaminating vector sequences or nonessential flanking chromosomal sequences.
5. This laboratory prefers to use Biotrans™ nylon membranes (ICN, Costa Mesa, CA). They are stronger and more resistant to drying and curling during the baking process. The maximum amount of DNA loaded in these membranes is approx 11 mg/mm^2. The user should calculate the area of the slot blot or dot blot apparatus being used and load DNA accordingly.
6. Hybridization temperatures cited are for *bphA*. The user should optimize and confirm all hybridization temperatures for different probes that may be used.

7. The hybridization buffer will solidify at room temperature. Before use, the user may have to heat gently to redissolve the SDS. After the membranes are removed from the heat sealable bags, allow the hybridization buffer to solidify and dispose of in solid ^{32}P waste.
8. Quantitation of hybrids can be accomplished by imaging software if you are using X-ray film. Insertion of sample intensity values into the standard curve regression equation of programs such as SigmaGel (Jandel Scientific, San Rafael, CA) can produce values of DNA hybridized to the probe. Direct quantification of hybridization signal can be performed with phosphor imaging or other beta counting equipment. Associated software will be able to generate the standard curve and fit the unknown data to the standard curve.

References

1. Hooper, S. W., Pettigrew, C. A. and Sayler, G. S. (1990) Ecological fate, effects, and prospects for the elimination of environmental polychlorinated biphenyls (PCBs). *Environ. Toxicol. Chem.* **9**, 655–667.
2. Cogliano, V. J. (1998) Assessing the cancer risk from environmental PCBs. *Environ. Health Perspect.* **106,** 317–323.
3. Ahmed, M. and Focht, D. D. (1973) Degradation of polychlorinated biphenyls by two species of *Achromobacter*. *Can. J. Microbiol.* **19,** 47–52.
4. Harkness, M. R., McDermott, J. B., Abramowicz, D. A., Salvo, J. J., Flanagan, W. P., Stephens, M. L., et al. (1993) In situ stimulation of aerobic PCB biodegradation in Hudson River sediments. *Science* **259,** 503–507.
5. Bedard, D. L. and Quensen, J. F., III. (1995) Microbial reductive dechlorination of polychlorinated biphenyls. In *Microbial Transformation and Degradation of Toxic Organic Chemicals* (Young, L, and Cerniglia, C., eds.), Wiley-Liss, Inc., New York, pp. 127–216.
6. Sayler, G. S., Sherrill, T. W., Perkins, R. E., Mallory, L. M., Shiaris, M. P., and Pedersen, D. (1982) Impact of coal-coking effluent on sediment microbial communities: a multivariate approach. *Appl. Environ. Microbiol.* **44,** 1118–1129.
7. Sayler, G. S., Perkins, R. E., Sherrill, T. W., Perkins, B. K. Reid, M. C., Shields, M. S., et al. (1983) Microcosm and experimental pond evaluation of microbial community response to synthetic oil contamination in freshwater sediments. *Appl. Environ. Microbiol.* **46,** 211–219.
8. Mallory, L. M. and Sayler, G. S. (1983) Heterotrophic bacterial guild structure: relationship to biodegradative populations. *Microbial Ecol.* **9,** 41–55.
9. Furukawa, K. (1994) Molecular genetics and evolutionary relationship of PCB-degrading bacteria. *Biodegradation* **5,** 289–300.
10. Bartels, F., Backhaus, S., Moore, E. R., Timmis, K. N., and Hofer, B. (1999) Occurrence and expression of glutathione-S-transferase-encoding *bphK* genes in *Burkholderia* sp. strain LB400 and other biphenyl-utilizing bacteria. *Microbiology* **145,** 2821–2834.
11. Hofer, B. Backhaus, S., and Timmis, K. N. (1994) The biphenyl/polychlorinated

biphenyl-degradation locus (*bph*) of *Pseudomonas* sp. LB400 encodes four additional metabolic enzymes. *Gene* **144,** 9–16.
12. Erickson, B. D. and Mondello, F. J. (1992) Nucleotide sequencing and transcriptional mapping of genes encoding biphenyl dioxygenase, a multicomponent PCB-degrading enzyme in *Pseudomonas* strain LB400. *J. Bacteriol.* **174,** 2903–2912.
13. Hofer, B., Eltis, L. D., Dowling, D. N., and Timmis, K. N. (1993) Genetic analysis of a *Pseudomonas* locus encoding a pathway for biphenyl/polychlorinated biphenyl degradation. *Gene* **130,** 47–55.
14. Furukawa, K. and Miyazaki, T. (1986) Cloning of a gene cluster encoding biphenyl and chlorobiphenyl degradation in *Pseudomonas pseudoalcaligenes. J. Bacteriol.* **166,** 392–398.
15. Taira, K., Hirose, J., Hayashida, S., and Furukawa, K. (1992) Analysis of *bph* operon from the polychlorinated biphenyl-degrading strain of *Pseudomonas pseudoalcaligenes* KF707. *J. Biol. Chem.* **267**, 4844–4853.
16. Kahn, A. and Walia, S. (1989) Cloning of bacterial genes specifying degradation of 4-chlorobiphenyl from *Pseudomonas putida* OU83. *Appl. Environ. Microbiol.* **55,** 798–805.
17. Walia, S., Khan, A., and Rosenthal, N. (1990) Construction and applications of DNA probes for detection of polychlorinated biphenyl-degrading genotypes in toxic organic-contaminated soil environments. *Appl. Environ. Microbiol.* **56,** 254–259.
18. Kahn, A. A., Wang, R. F., Nawaz, M. S., and Cerniglia, C. E. (1997) Nucleotide sequence of the genes encoding cis-biphenyl dihydrodiol dehydrogenase (*bph*B) and the expression of an active recombinant His-tagged *bphB* gene product from PCB degrading bacterium, *Pseudomonas putida* OU83. *FEMS Microbiol. Lett.* **154,** 317–324.
19. Ahmad, D., Masse, R., and Sylvestre, M. (1990) Cloning and expression of genes involved in 4-chlorobiphenyl transformation by *Pseudomonas testosteroni*: homology to polychlorobiphenyl-degrading genes in other bacteria. *Gene* **86,** 53–61.
20. Ahmad, D., Sylvestre, M., and Sondossi, M. (1991) Subcloning of bph genes from *Pseudomonas testosteroni* B-356 in *Pseudomonas putida* and *Escherichia coli*: evidence for dehalogenation during initial attack on chlorobiphenyls. *Appl. Environ. Microbiol.* **57,** 2880–2887.
21. Sylvestre, M., Sirois, M., Hurtubise, Y., Bergeron, J., Ahmad, D., Shareck, F., et al. (1996) Sequencing of *Comamonas testosteroni* strain B-356 biphenyl/chlorobiphenyl dioxygenase genes: Evolutionary relationships among Gram-negative bacterial biphenyl dioxygenases. *Gene* **174,** 195–205.
22. Sharma, A., Chunn, C. D., Rothmel, R. K., and Unterman, R. (1991) Studies on bacterial degradation of polychlorinated biphenyls: optimization of parameters for *in vivo* enzyme activity. Poster Q48. *91st Gen. Meet Am. Soc. Microbiol.*, May 5–9, Dallas, TX.
23. Asturias, J. A., Diaz, E., and Timmis, K. N. (1995) The evolutionary relationship of biphenyl dioxygenase from gram-positive *Rhodococcus globerulus* P6 to multicomponent dioxygenease from Gram-negative bacteria. *Gene* **156,** 11–18.

24. Yamada, A., Kishi, H., Sugiyama, K., Hatta, T., Nakamura, K. Masai, E., and Fukuda, M. (1998) Two nearly identical aromatic compound hydrolase genes in a strong polychlorinated biphenyl degrader, *Rhodococcus* sp. strain RHA1. *Appl. Environ. Microbiol.* **64,** 2006–2012.
25. Masai, E., Yamada, A., Healy, J. M., Hatta, T., Kimbara, K., Fukuda, M., and Yano, K. (1995) Characterization of biphenyl catabolic genes of Gram-positive polychlorinated biphenyl degrader *Rhodococcus* sp. strain RHA1. *Appl. Environ. Microbiol.* **61,** 2079–2085.
26. Masai, E., Sugiyama, K., Iwashita, N., Shimizu, S., Hauschild, J. E., Hatta, T., et al. (1997) The *bph*DEF *meta*-cleavage pathway genes involved in biphenyl/ polychlorinated biphenyl degradation are located on a linear plasmid and separated from the initial *bph*ACB genes in *Rhodococcus* sp. strain RHA1. *Gene* **187,** 141–149.
27. Hauschild, J. E., Masai, E., Sugiyama, K., Hatta, T., Kimbara, K., Fukuda, M., and Yano, K. (1996) Identification of an alternative 2,3-dihydroxybiphenyl 1,2-dioxygenase in *Rhodococcus* sp. strain RHA1 and cloning of the gene. *Appl. Environ. Microbiol.* **62,** 2940–2946.
28. Kosono, S., Maeda, M., Fuji, F., Arai, H., and Kudo, T. (1997) Three of the seven *bph*C genes of *Rhodococcus erythropolis* TA421, isolated from a termite ecosystem, are located on an indigenous plasmid associated with biphenyl degradation. *Appl. Environ. Microbiol.* **63,** 3282–3285.
29. Shimura, M., Mukerjee-Dhar, G., Kimbara, K., Nagato, H., Kiyohara, and Hatta, T. (1999) Isolation and characterization of a thermophilic *Bacillus* sp. JF8 capable of degrading polychlorinated biphenyls and naphthalenes. *FEMS Microbiology Letters* **178,** 87–93.
30. Bedard, D., Untermann, R., Bopp, L. H., Brennan, M. J., Haberl, M. L., and Johnson, C. (1986) Rapid assay for screening and characterizing microorganisms for the ability to degrade polychlorinated biphenyls. *Appl. Environ. Microbiol.* **51,** 761–768.
31. Mondello, F. J., Turcich, M. P., Lobos, J. H., and Erickson, B. D. (1997) Identification and modification of biphenyl dioxygenase sequences that determine the specificity of polychlorinated biphenyl degradation. *Appl. Environ.Microbiol.* **63,** 3096–3103.
32. Seto, M., Kimbara, K., Shimura, M., Hatta, T., Fukuda, M., and Yano, K. (1995) A novel transformation of polychlorinated biphenyls by *Rhodococcus* sp. strain RHA1. *Appl. Environ. Microbiol.* **61,** 3353–3358.
33. Peng, X., Egashira, T., Hanashiro, K., Masai, E., Nishikawa, S., Katayama, et al. (1998) Cloning of a *Sphingomonas paucimobilis* SYK-6 gene encoding a novel oxygenase that cleaves lignin-related biphenyl and characterization of the enzyme. *Appl. Environ. Microbiol.* **64,** 2520–2527.
34. Yadav, J. S., Quensen, J. F., III, Tiedje, J. M., and Reddy, C. A. (1995) Degradation of polychlorinated biphenyl mixtures (Aroclors 1242, 1254, and 1260) by the white rot fungus *Phanerochaete chyrosporium* as evidenced by congener-specific analysis. *Appl. Environ. Microbiol.* **61,** 2560–2565.
35. Erb, R.W. and Wagner-Dobbler, I. (1993) Detection of polychlorinated biphenyl

degradation genes in polluted sediments by direct DNA extraction and polymerase chain reaction. *Appl. Environ. Microbiol.* **59,** 4065–4073.

36. Layton, A. C., Lajoie, C. A., Easter, J. P., Jernigan, R., Sanseverino, J., and Sayler. G. S. (1994) Molecular diagnostics and chemical analysis for assessing biodegradation of polychlorinated biphenyls in contaminated soils. *J. Industr. Microbiol.* 13, 392–401.
37. Pellizari, V. H., Bezborodnikov, S., Quensen, J. F.,III, and Tiedje, J. M. (1996) Evaluation of strains isolated by growth on naphthalene and biphenyl for hybridization of genes to dioxygnease probes and polychlorinated biphenyl-degrading ability. *Appl. Environ. Microbiol.* **62,** 2053–2058.
38. Van Dyke, M. I., Lee, H., and Trevors, J. T. (1996) Survival of *lux*AB marked *Alcaligenes eutrophus* H850 in PCB-contaminated soil and sediment. *J. Chem. Tech. Biotech.* **65,** 115–122.
39. Kimura, N., Nishi, A., Goto, M., and Furukawa, K. (1997) Functional analyses of a variety of chimeric dioxygenases constructed from two biphenyl dioxygenases that are similar structurally but different functionally. *J. Bacteriol.* **179,** 3936–3943.
40. Anderson, L. M. and Young, B. D. (1985) Quantitative filter hybridization. In *Nucleic Acid Hybridization: A Practical Approach* (Hames, B. D. and Higgins, S. J., eds.), IRL Press Limited, Oxford, England, Chapter 4, pp. 73-111.
41. Bedard, D., Wagner, R. E., Brennan, M. J., Haberl, M. L., and Brown, J. F., Jr. (1987) Extensive degradation of aroclors and environmentally transformed polychlorinated biphenyls by *Alcaligenes eutrophus* H850. *Appl. Environ. Microbiol.* **53,** 1094–1102.
42. Thompson, J. D., Higgins, D. G., and Gibson, T. J. (1994) Clustal W: improving sensitivity of progressive multiple sequence alignment through sequence weighting, position-specific gap penalties and weight matrix choice. *Nucleic Acids Res.* **22,** 4673–4680.
43. Page, R. D. (1996) TREEVIEW: an application to display phylogenetic trees on personal computers. *Compu. Appl. Biosci.* **12,** 357–359.
44. Church, G. M. and Gilbert, W. (1984) Genomic sequencing. *Proc. Natl. Acad. Sci. USA* **81,** 1991–1995.
45. Ogram, A., Sayler, G. S., and Barkay, T. (1987) The extraction and purification of microbial DNA from sediments. *J. Microbiol. Methods* **7,** 57–66.
46. Holben, W. E., Jansson, J. K., Chelm, B. K., and Tiedje, J. M. (1988) DNA probe method for the detection of specific microorganisms in the soil bacterial community. *Appl. Environ. Microbiol.* **54,** 703–711.
47. Tsai, Y.-L. and Olson, B. H. (1991) Rapid method for direct extraction of DNA from soil and sediments. *Appl. Environ. Microbiol.* **57,** 1070–1074.
48. Picard, C., Ponsonnet, C., Paget, E., Nesme, X., and Simonet, P. (1992) Detection and enumeration of bacteria in soil by direct DNA extraction and polymerase chain reaction. *Appl. Environ. Microbiol.* **58,** 2717–2722.
49. Jacobsen, C. S. and Rasmussen, O. F. (1992) Development and application of a

new method to extract bacterial DNA from soil based on separation of bacteria from soil with cation-exchange resin. *Appl. Environ. Microbiol.* **58,** 2458–2462.

50. Stapleton, R. D., Ripp, S., Jimenez, L. Cheol-Koh, S., Fleming, J. T., Gregory, I. R., and Sayler, G. S. (1998) Nucleic acid analytical approaches in bioremediation: site assessment and characterization. *J. Microbiol. Methods.* **32,** 165–178.
51. Stapleton, R. D. (1997) Natural attenuation of petroleum hydrocarbons in groundwater. Ph.D. dissertation. University of Tennessee, Knoxville, USA.
52. Promega (1992) *Technical Manual 16*. Madison, Wisconsin. USA.
53. Applegate, B. M., Matrubutham, U., Sanseverino, J., and Sayler, G. S. (1995) Biodegradation genes as marker genes in microbial ecosystems. *Molecular Microbial Ecology Manual* **6.1.8,** 1–14.
54. Johnston, W. H., Stapleton, R., and Sayler, G. S. (1996) Direct extraction of microbial DNA from soils and sediments. *Molecular Microbial Ecology Manual* **1.3.2**, 1–9.
55. Stahl, D. A., Flesher, B., Mansfield, H. R., and Montgomery, L. (1988) Use of phylogenetically-based hybridization probes for studies of ruminal microbial ecology. *Appl. Environ. Microbiol.* **54,** 1079–1084.
56. Smith, C. L., Econome, J. G., Schutt, A., Klco, S., and Cantor, C. R. (1987) A physical map of the *Escherichia coli* K12 genome. *Science* **236,** 1448–1453.
57. Klappenbach, J. A., Dunbar, J. M. and Schmidt, T. M. (2000) rRNA operon copy number reflects ecological strategies of bacteria. *Appl. Environ. Microbiol.* **66,** 1328–1333.

20

Detection of Single-Copy Genes in DNA from Transgenic Plants

Lee C. Garratt, Matthew S. McCabe, J. Brian Power, and Michael R. Davey

1. Introduction

The development of increasingly sensitive chemiluminescent substrates and the ability to label probes with digoxigenin (DIG) by polymerase chain reaction (PCR) *(1)* has resulted in nonradioactive Southern analysis becoming the preferred method, in many plant research laboratories, for the detection of single-copy genes in DNA from transgenic plants. The previous, well established procedure for single-copy gene detection required the utilization of ^{32}P-labeled probes. However, as well as safety issues, isotopic probes require labeling immediately prior to use with exposure times ranging from 1 d to 7 d. Isotopic probes are therefore laborious and time consuming. The procedure outlined here is rapid and simple and employs modification of two published protocols *(1,2)*, using DIG-labeled nucleotides *(3)* that are incorporated into nucleic acid probes by PCR *(1,4,5)*. The analysis of a fragment of dissected transgene, attached to plant genomic DNA (border fragment analysis), gives information on transgene integrity and integration pattern into plant DNA, as well as transgene copy number. The number of bands, following chemiluminescent detection, corresponds to the number of transgene copies (*see* **Fig. 1**). The analysis of either the whole or part of the transgene, which has been dissected from the plant genomic DNA using restriction enzymes (internal fragment analysis), provides information on transgene copy number and integrity, but not the integration pattern. Following chemiluminescent detection, an evaluation of band intensity, relative to the band intensity of known single-copy and multiple-copy T-DNA inserts, also allows transgene copy number to be determined.

From: *Methods in Molecular Biology, vol. 179: Gene Probes: Principles and Protocols*
Edited by: M. Aquino de Muro and R. Rapley © Humana Press Inc., Totowa, NJ

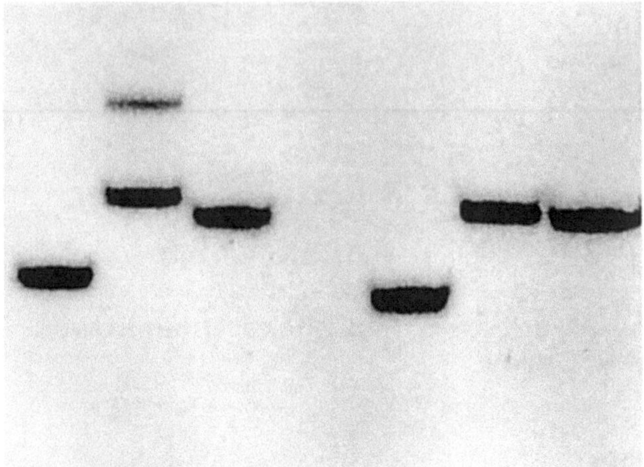

Fig. 1. Southern blot of DNA from transgenic lettuce plants (lanes 1–3 and 5–7) digested with *Hin*dIII to produce border fragments, followed by hybridization with a PCR-DIG-labeled *luc* (luciferase reporter gene) probe. Lanes 1, 3, 5, 6, and 7 represent transgenic plants with single-copy gene inserts; lane 2 is a transgenic plant containing two gene inserts, while lane 4, in which the sample does not hybridize to the PCR-DIG labeled *luc* probe, represents DNA derived from a nontransformed lettuce plant.

2. Materials

Chemicals (reagent grade) are from Sigma-Aldrich, Poole, UK, unless stated.

2.1. Extraction of Plant DNA for Southern Analysis

1. Extraction buffer: 100 m*M* Tris-HCl (12.11 g/L of Trizma base, pH 8.0), 50 m*M* EDTA (18.61 g/L of EDTA-disodium salt), 500 m*M* NaCl (29.22 g/L), 10 m*M* β-mercaptoethanol (700 µl/L, *see* **Note** 1). Autoclave for 20 min at 121°C.
2. Sodium dodecyl sulfate (SDS) solution (200 g/L; *see* **Note 2**).
3. 5 *M* Potassium acetate (490.70 g/L).
4. Resuspension buffer: 50 m*M* Tris-HCl (6.05 g/L of Trizma base, pH 8.0), 10 m*M* EDTA (3.72 g/L of EDTA-disodium salt). Autoclave as described in **Subheading 2.1.1**.
5. CTAB buffer: 200 m*M* Tris-HCl (24.22 g/L of Trizma base, pH 7.5–8.0), 50 m*M* EDTA (18.61 g/L of EDTA-disodium salt), 2 *M* NaCl (116.88 g/L), 2% CTAB (20 g/L hexadecyltrimethylammonium bromide). Autoclave as described in **Subheading 2.1. item 1.**

6. TE buffer: 10 m*M* Tris-HCl (1.21 g/L of Trizma base, pH 8.0), 1 m*M* EDTA (372.20 mg/L of EDTA-disodium salt).
7. Polypropylene screw capped tubes (50-mL capacity) (Alpha Laboratories, Eastleigh, UK).
8. Water bath at 65°C.
9. Miracloth (Calbiochem, Nottingham, UK).
10. Phenol–chloroform–isoamyl alcohol (24:23:1 by volume) (*see* **Note 3**).
11. Isopropanol.
12. 3 *M* Sodium acetate (408.30 g/L) adjusted to pH 5.2 with glacial acetic acid.
13. Ethanol, 70% aqueous solution.
14. RNase A (16 mg/mL) (70 Kunitz units/mg, EC 3.1.27.5; Sigma reference R 4642).
15. Chloroform–isoamyl alcohol (24:1 by vol).

2.2. DNA Quantification for Southern Analysis

1. Hoechst dye (*Bis*-benzimide fluorochrome trihydrochloride [Calbiochem, Nottingham, UK; cat. no. 382061]) (*see* **Note 4**).
2. 10× TNE buffer: 100 μ*M* Tris-HCl (12.10 g/L of Trizma base, pH 7.4), 1 *M* NaCl (58.40 g/L), 10 μ*M* EDTA (3.70 g/L of EDTA-disodium salt); working dilution, 1× TNE.
3. Microtiter plates (Nunclon™, Kamstrupvej, Denmark).
4. Fluorescence plate reader (Cytofluor 2300; Millipore, Watford, UK).
5. Calf thymus DNA type 1: 5 mg/10 mL.
6. TE buffer: 10 m*M* Tris-HCl (1.21 g/L of Trizma base, pH 8.0), 1 m*M* EDTA (372.20 mg/L of EDTA-disodium salt).

2.3. Sample Preparation for Gel Electrophoresis

1. 5× TBE buffer: 0.45 *M* Trizma base (54.50 g/L of Trizma base, pH 8.0), 0.45 *M* boric acid (27.50 g/L), 10 m*M* EDTA (3.70 g/L of EDTA-disodium salt).
2. DNA loading buffer: 15% (w/v) Ficoll (Type 400 nonionic synthetic polymer), 0.25% (w/v) bromophenol blue, 0.25% (w/v) xylene cyanol FF (*see* **Note 5**).
3. 3 *M* Sodium acetate (408.30 g/L), adjusted to pH 5.2 with glacial acetic acid.
4. Isopropanol.
5. Absolute ethanol.
6. Ethanol, 70% aqueous solution.
7. Appropriate restriction enzyme(s) (*see* **Note 6**).
8. 10× dilution buffer (i.e., SuRE/Cut buffer B; Boehringer Mannheim, Lewes, UK; cat. no. 1417967) (*see* **Note 6**).
9. Agarose, molecular biology grade (New Brunswick Scientific, Hatfield, UK, or Flowgen Instruments, Sittingbourne, UK).
10. 8% (w/v) TBE gel: 300 mL of 1× TBE (1:5 dilution of 5× TBE), 2.40 g of agarose. Heat until dissolved. Add 3 μL of ethidium bromide solution (10 mg/mL) 2 min after agarose has dissolved; mix and allow gel to set.

11. DIG-labeled DNA molecular weight marker V (Boehringer Mannheim; cat. no. 1669931).
12. Electrophoresis apparatus for 300-mL horizontal agarose gels.
13. UV transilluminator (Chromato-VUE, model TM-20; UVP, San Gabriel, USA).
14. Vacuum desiccator (Kartell, SLS, Nottingham, UK).

2.4. Alkaline Transfer (Capillary Blotting)

1. Positively charged nylon membrane (Boehringer Mannheim; cat. no. 1417240).
2. Transfer buffer: 0.4 *M* NaOH (16 g/L), prepared immediately before use.
3. Whatman 3MM filter paper (Whatman Laboratory Division, Maidstone, UK).
4. Saran Wrap^{TM} (Dow Chemical Company, Kings Lynn, UK) or domestic Cling Film.
5. Capillary blot transfer apparatus ***(6)*** (*see* **Note 7**).
6. 20× SSC: 300 m*M* NaCl (175.30 g/L), 30 m*M* sodium citrate (88.20 g/L), pH 7.0. Autoclave as described in **Subheading 2.1. item 1**.
7. 2× Saline sodium citrate (SSC): 100 mL of 20× SSC in 900 mL of dH_2O.
8. Oven at 120°C.

2.5. PCR DIG-Probe Labeling

1. Thermal cycler (Genius, Techne, Cambridge, UK).
2. PCR DIG-labeling mixture: 2 m*M* dATP, 2 m*M* dCTP, 2 m*M* dGTP, 1.3 m*M* dTTP, 0.7 m*M* DIG-11-dUTP, and *Taq* expand high-fidelity DNA polymerase (Boehringer Mannheim). Alternatively, a PCR DIG-labeling kit is available from Boehringer Mannheim (cat. no. 1636090).
3. TE buffer: 10 m*M* Tris-HCl (1.21 g/L of Trizma base, pH 8.0), 1 m*M* EDTA (372.20 mg/L of EDTA-disodium salt).
4. 10× PCR buffer: 20 m*M* $(NH)_2SO_4$ (2.64 g/L), 750 m*M* Tris-HCl (907.50 g/L, pH 8.0), 0.1% Tween-20 (polyoxyethylene-sorbitan monolaurate), pH 9.0.
5. Reverse and forward primers for DNA sequence of interest (*see* **Note 8**).
6. Template DNA containing the DNA sequence of interest.
7. Nanopure or milli-Q water. Autoclave as described in **Subheading 2.1. item 1**.
8. 3 *M* sodium acetate (408.30 g/L), adjusted to pH 5.2 with glacial acetic acid.
9. Absolute ethanol.
10. Ethanol, 70% aqueous solution.
11. Ice.
12. Agarose, molecular biology grade (New Brunswick Scientific or Flowgen Instruments Ltd.).
13. 1.5% (w/v) TBE gel: 300 mL 1× TBE (1:5 dilution of 5× TBE), 4.50 g of agarose. Prepare as described in **Subheading 2.3.10**.

2.6. Prehybridization and Hybridization

1. DIG Easy Hyb solution (Boehringer Mannheim; cat. no. 1603558).
2. Hybridization oven with Rotisserie, 37°C (e.g., Model SI 20H, Stuart Scientific, Stone, UK).

3. Water bath at 95°C.
4. Ice.

2.7. Posthybridization Washing

1. Wash solution A: 2× SSC (25 mL of 20× SSC, 100 mL of dH_2O), 0.1% SDS (1.25 mL of 20% w/v SDS). Add 25 mL of 20× SSC to 100 mL of dH_2O BEFORE adding SDS; then make up to 250 mL with dH_2O. Prepare immediately prior to use.
2. Wash solution B: Same as wash solution A, except 25 mL 20 × SSC is replaced with 6.25 mL of 20× SSC. Make up to 250 mL with dH_2O. Prepare immediately before use.
3. Hybridization oven at 65°C with Rotisserie.
4. Water bath at 65°C.

2.8. Chemiluminescent Detection

1. Buffer A: 0.1 *M* maleic acid (11.60 g/L), 3 *M* NaCl (175.30 g/L), 0.2 *M* NaOH (8.00 g/L). Make up to 800 mL with dH_2O and heat the solution at 60°C while stirring to dissolve the NaCl; adjust to pH 8.0 with 1.0 *M* and 0.1 *M* NaOH solutions. Adjust to 1 L with dH_2O. Autoclave as described in **Subheading 2.1.1.** and allow to cool. Add 3 mL of Tween-20.
2. Buffer B: Same as buffer A, but with 0.15 *M* NaCl (8.80 gL).
3. 5% Blocking reagent buffer: 5 g of Blocking reagent (Boehringer Mannheim; cat. no. 1096176), 100 mL of buffer solution B. Dissolve at 60–90°C, but do not allow to boil until fully dissolved (*see* **Note 9**). Autoclave as described in **Subheading 2.1, item 1**.
4. Buffer C: 20 mL of 5% blocking reagent buffer, 80 mL of buffer A. Prepare immediately before use.
5. Buffer D: 0.1 *M* Tris-HCl (12.10 g of Trizma base, pH 9.5), 0.1 *M* NaCl (5.84 g L).
6. Substrate solution A: 5 mL of buffer D, 50 µL of CPD-Star Chemiluminescence substrate (Boehringer Mannheim; cat. no. 1685627).
7. Developer: 100 mL of LX-24 X-ray developer (Kodak, Hemel Hempstead, UK), 400 mL of dH_2O.
8. Fixer: 100 mL of Rapid Fixer (Ilford, Mobbereley, UK), 400 mL of distilled water.
9. X-ray film: Hyperfilm™ X-ray film (Amersham Life Science, Little Chalfont, UK).
10. Anti-DIG-AP: Polyclonal sheep anti-digoxigenin Fab fragments, conjugated to alkaline phosphatase (Boehringer Mannheim; cat. no. 1093274).
11. Glass casserole dish or plastic box large enough to allow filters to lie flat without upturned corners. These vessels must be cleaned thoroughly before use.
12. X-ray film cassette.

2.9. Stripping and Reprobing

1. Lauryl sulfate (SDS) (1.0 g/L).
2. Glass container, large enough to allow membrane to lie flat and to be covered by 1 L of solution.

3. Methods

3.1. DNA Extraction for Southern Analysis

1. Freeze 0.5–0.75 g of plant material in liquid N_2. Grind to a fine powder using a cooled pestle and mortar. Add liquid N_2 to keep the material frozen and brittle.
2. Transfer the powder to a 50-mL polypropylene tube containing 15 mL of extraction buffer.
3. Add 1 mL of 20% SDS and mix thoroughly by shaking. Incubate tube in a water bath at 65°C for 10 min.
4. Add 5 mL of 5 *M* potassium acetate and mix thoroughly by shaking. Incubate the tube at 4°C for at least 20 min.
5. Centrifuge the tube for 10 min (3000*g*). Pour the supernatant through "Miracloth" into a clean 50-mL tube containing 10 mL of isopropanol (*see* **Note 10**).
6. Mix by inverting the tube and incubate at –20°C for 20 min. Centrifuge the tube for 10 min (3000*g*). Decant the supernatant and allow the pellet to drain. Redissolve the pellet in 0.7 mL of resuspension buffer and transfer to a 1.5-mL microfuge tube. Microfuge for 10 min (3000*g*) to pellet insoluble debris (*see* **Note 11**).
7. Transfer the supernatant to a clean microfuge tube, add an equal volume phenol–chloroform–isoamyl alcohol of (24:23:1 by vol). Mix by shaking and microfuge for 10 min (3000*g*).
8. Transfer the aqueous phase (600 µL) to a new microfuge tube. Add 0.1 volume (60 µL) of 3 *M* sodium acetate and 0.6 volume (360 µL) of isopropanol.
9. Mix thoroughly by inversion (*see* **Note 12**) and incubate at –20°C for 10 min. Mix again by inverting the tube; microfuge for 10 min (3000*g*).
10. Discard the supernatant and wash the pellet with 600 µL of cold 70% ethanol. Microfuge for 2 min (3000*g*).
11. Discard the supernatant and microfuge for 30 s (3000*g*). Remove the remaining ethanol with a micropipet (all the ethanol MUST be removed).
12. Dry the DNA pellet and redissolve in 350 µL of TE buffer or dH_2O.
13. Add 6 µL of RNase A (16 mg/mL) and incubate in a water bath at 37°C for 15 min.
14. Add an equal volume of CTAB buffer; incubate in a water bath at 60°C for 15 min.
15. Add 700 µL of cloroform–isoamyl alcohol (24:1 by vol) and mix by shaking until an emulsion is formed (*see* **Note 13**).
16. Microfuge for 10 min (3000*g*).
17. Transfer the aqueous phase (600 µL) to a clean microfuge tube; add 360 µL of isopropanol and mix by inversion.

18. Microfuge for 10 min (3000*g*).
19. Discard the supernatant and wash the pellet with 600 of μL cold 70% ethanol. Microfuge for 2 min (3000*g*).
20. Discard the supernatant and microfuge for 30 s (3000*g*). Remove the remaining ethanol with a micropipet (all the ethanol MUST be removed).
21. Dry the pellet for 5 min (do not overdry) and re-suspend (*see* **Note 14**) in 100 μL of distilled water.
22. Resuspended DNA can be stored at 4°C for up to 4 wk, or at –20°C for 12 mo.

3.2. DNA Quantification for Southern Analysis

1. Add 6 μL of resuspended DNA to 100 μL of Hoechst dye (0.5 μg/mL in 1× TNE buffer) to each of three wells of a microtiter plate, creating three replicates per sample.
2. Mix the sample(s) by gently agitating the microtiter plate.
3. Cover the microtiter plate with aluminum foil prior to reading to protect the samples from light.
4. Standards of 0.5-, 1.0-, 1.5-, 2.0-, and 2.5-μg aliquots of calf thymus DNA type I should be made up to 6 μL with TE buffer. Add 200 μL of Hoechst dye (0.5 μg/mL in 1× TNE). Add each standard to a well on the microtiter plate.
5. Quantify the reaction using a fluorescence plate reader at 360 nm with emission at 460 nm (*see* **Note 15**).

3.3. Sample Preparation for Electrophoresis

1. After quantifying the DNA of each sample, add the respective volume required for 10 μg of DNA to a clean microfuge tube.
2. To 10 μg of DNA add 20 μL of 10× dilution buffer and 50 U of restriction enzyme (*see* **Note 16**); adjust to 200 μL with distilled water.
3. Incubate samples for 16–48 h (not more than 48 h) at 37°C; maintain, if necessary, at 4°C.
4. Add 300 μL of dH_2O, 50 μL of 3 *M* sodium acetate, and 0.6 vol (180 μL) of isopropanol; mix by inversion.
5. Incubate at –20°C for 10 min.
6. Microfuge for 10 min (3000g) and retain the pellet.
7. Wash the pellet with 600 μL of cold 70% ethanol. Microfuge for 2 min (3000*g*). Discard the supernatant and again microfuge for 30 s. Remove the remaining ethanol with a micropipet (all the ethanol MUST be removed).
8. Dry the pellet in a vacuum desiccator (*see* **Note 17**).
9. Resuspend the pellet in 12 μL of dH_2O_2 and add 3 μL of loading dye (*see* **Note 18**).
10. Incubate the samples in a water bath at 60°C for 5 min; place on ice for 2–5 min.
11. Run samples on a 300 μL of 0.8% (w/v) agarose TBE gel for 16 h at 25 V (*see* **Note 19**). Also load 10 μL of DIG-labeled DNA molecular weight maker V, mixed with 2 μL of loading dye, at least three wells apart from other samples (*see* **Note 20**).

12. Place the gel (still in the casting tray) in a tray containing 2 L of distilled water and 100 µL of ethidium bromide (1.0 mg/mL). Leave for 0.5 h, agitating occasionally.
13. Assess DNA digestion by observation and/or photography on a UV transilluminator. Do not expose the gel to UV for more than 30 s to avoid damaging the DNA.

3.4. Alkaline Transfer (Capillary Blotting)

1. Immerse the gel, on the casting tray, in 0.4 *M* NaOH for 15 min. Agitate by hand every 2 min.
2. Set up the capillary blot *(6)*. Place the gel support in the blotting tank. Fold a sheet of Whatman 3MM filter paper over the gel support so that two ends of the filter paper rest in the blotting tank. Moisten the filter paper with transfer buffer (0.4 *M* NaOH) and add 2 L of transfer buffer to the blotting tank. Place the gel, loading side down, on the filter paper and overlay the gel with a positively charged nylon membrane (Boehringer Mannheim) cut to the size of the gel (*see* **Note 21**). Overlay the membrane with 2× Whatman 3MM filter papers cut to size, and 6 cm of absorbent towels. Overlay the assembly with a glass sheet and a 500-g weight (e.g., a 500-mL medical flat bottle filled with water) and blot for 16 h.
3. Wash/neutralize the membrane in 2× SSC (the dye front should change to green).
4. Place in an oven and heat at 120°C for 0.5 h.
5. If storage is required, wrap the membrane in SaranWrap™ or Cling Film and store flat at –20°C.

3.5. PCR DIG-Probe Labeling

1. Add to a sterile microfuge tube on ice, 5 µL of 10× PCR buffer, 1.0–5.0 m*M* $MgCl_2$, 5 µL of 10× PCR DIG-labeling mixture, 0.1–1.0 µ*M* forward primer, 0.1–1.0 µ*M* reverse primer, 0.5–2.5 U *Taq* DNA polymerase, and the required volume of template DNA (*see* **Note 22**).
2. Mix the PCR reaction components and microfuge for 10 s (3000*g*).
3. Use the optimum cycling conditions **(*7*)** for the primers and template DNA to amplify the sequence of interest.
4. Following amplification, run 5 µL of the PCR product on a 1.5% (w/v) TBE agarose gel to check amplification (*see* **Note 22**).
5. To 42 µL of the PCR product, add 4.2 µL (0.1 volume) of 3 *M* sodium acetate and 105 µL (2.5 volume) of absolute ethanol. Invert the tube to precipitate the DNA.
6. Incubate at –80°C for 10 min.
7. Microfuge for 10 min (3000*g*), to pellet the DNA; discard the supernatant.
8. Wash the pellet with ice-cold 70% ethanol and microfuge for 2 min (3000*g*).
9. Discard the supernatant and microfuge for 30 s (3000*g*). Remove the remaining ethanol from above the DNA pellet with a micropipet (all the ethanol MUST be removed).
10. Resuspend the DNA pellet in 50 µL of TE buffer.
11. Store the DIG-labeled DNA probe at –20°C.

3.6. Prehybridization and Hybridization

1. While handling only the edges of the nylon membrane carrying the blotted DNA (*see* **Subheading 3.4.5.**), roll the membrane and place it in a Rotisserie tube, avoiding as much overlap as possible. The back of the membrane must be in contact with the walls of the tube.
2. Prewarm 20 mL of DIG Easy Hyb solution to 37°C in a water bath (*see* **Note 23**) and add to the Rotisserie tube containing the membrane. Prehybridize the membrane at 37°C in an oven for 1 h at 4 rpm (*see* **Note 24**).
3. Denature the DIG-labeled probe (*see* **Subheading 3.5.11.**) by heating above 95°C for 10 min in a water bath; cool on ice for 5 min.
4. Prepare the hybridization solution by adding 25 μL of denatured DIG-labeled probe to 10 mL of DIG Easy Hyb solution; mix thoroughly (*see* **Note 25**).
5. Decant the prehybridization solution from the Rotisserie tube (the solution can be retained for future use) and replace with the hybridization solution.
6. Hybridize at 37°C, for 16 h at 4 rpm.
7. Decant the hybridization solution. After use, store the hybridization solution at –20°C for reuse three or four times.

3.7. Posthybridization Washing

1. Add 50 mL of wash solution A to the membrane in the Rotisserie tube and rotate at 4 rpm in the hybridization oven at room temperature for 5 min. Decant the solution; repeat with new solution.
2. Remove wash solution A and replace with 50 mL of wash solution B preheated to 65°C in a water bath. Incubate for 15 min in the hybridization oven at 65°C, 4 rpm (*see* **Note 26**).

3.8. Chemiluminescent Detection of Bound DIG-Probe

The following steps are performed at room temperature in a clean plastic box on a rocking shaker:

1. Rinse the membrane in buffer A for 1 min.
2. Block the membrane by incubating in 100 mL of buffer C for 1 h.
3. Microfuge (3000*g*) the anti-DIG-AP solution in its storage tube for 1 min to pellet any precipitate (*see* **Note 27**). Dilute (1:1000 by volume) 5 μL of anti-DIG-AP with buffer C (*see* **Note 28**).
4. Discard buffer C from the plastic box and add all of the anti-DIG-AP/buffer C solution. Incubate the membrane in the solution for exactly 30 min at room temperature.
5. Remove the unbound anti-DIG-AP by four washes (10 min each) with buffer A.
6. Remove buffer A and equilibrate the membrane for 5 min in 5 mL of buffer D.
7. Drain buffer D from the membrane, but do not allow the membrane to dry.
8. Seal the membrane between two sheets of Saran-Wrap™.
9. Expose the membrane to X-ray film in an X-ray cassette for 5–30 min.
10. Develop and fix the film.

3.9. Stripping and Reprobing

The following procedure can be performed if it proves necessary to remove bound probe from the DNA on the membrane (e.g., to permit hybridization with a different probe):

1. Heat 1 L of 0.1% (w/v) SDS solution to 95–98°C in a clean glass container large enough to allow the membrane to lie flat (*see* **Note 29**).
2. Immerse the membrane in the hot solution and incubate for exactly 10 min, while maintaining the temperature at 95–98°C.
3. Remove the membrane from the solution and drain, but do not allow the membrane to dry.
4. At this stage, prehybridization can be performed immediately (*see* **Subheading 3.6.**), or the membrane can be stored at –20°C, as described in **Subheading 3.4. step 5.**, until required.

4. Notes

1. β-mercaptoethanol should be used in a fume hood with forced ventilation. If used outside a fume hood, every precaution must be taken to minimize vapor release and to ensure adequate ventilation. Add β-mercaptoethanol to the extraction buffer immediately prior to use.
2. Lauryl sulfate (SDS) may precipitate from solution during use or storage, but can be redissolved by warming to 65°C in a water bath.
3. When purifying nucleic acids (i.e., removing proteins), ReadyRed™ (Appligene-Oncor; Rue Geiler de Kaysesberg, Illkirch, France) can be used instead of chloroform/isoamyl alcohol during phenolic extraction. The color of ReadyRed™ facilitates the visualization and separation of aqueous and organic phases, with the aqueous (top) phase appearing red.
4. Dissolve 20 mg in 20 mL of distilled water and prepare a 1:2000 by volume dilution by adding 10 μL of the solution to 20 mL of 1× TNE buffer.
5. Ficoll loading buffer is used in preference to glycerol-based loading buffers to increase the resolution of the DNA bands. Glycerol based buffers should be avoided since they interact with borate in TBE agarose gels, resulting in an alteration in the local pH in the gel.
6. Many restriction enzymes do not cut highly methylated DNA. In such cases, use enzymes, for example, *Xba*I, *Eco*RI, *Bam*HI, and *Bgl*II, which lack the methylation-sensitive nucleotide configuration (CNG) in their recognition sequence. Most restriction enzymes are supplied with their appropriate dilution buffer (for example, SuRE/Cut Buffer B from Boehringer Mannheim cat. no. 1417967). Alternately, the constituents of specific buffers for restriction endonuclease digestion can be found in **ref. *(6)***.
7. Vacuum blotting or electroblotting can be used as an alternative to capillary blot transfer, reducing the blotting time of DNA transfer from the gel to the membrane from 16 to 1–2 h. However, capillary blotting is preferable for DNA transfer to the membrane ***(1)***.

8. Primers should be designed to flank the region of interest. They should contain 40–60% G+C and care should be taken to avoid sequences that produce internal secondary structure. To avoid the production of primer-dimers in the PCR reaction, the 3' ends of the primers should not be complementary and, ideally, both primers should anneal at the same temperature. A range of software packages is available (e.g., OLIGO Primer Analysis; Lifescience Software, Longlake, USA) to assist in the design of the primers.
9. The solution should be straw colored without visible granules, as undissolved granules result in high background. The blocking reagent may be filtered through "Miracloth" to remove any granules.
10. Fold a 15-cm square of "Miracloth" to form a funnel and place the folded "Miracloth" in the top of a 50-mL tube. If a plug of plant material forms in the top of the sample tube, displace the plug to ensure a constant pour rate.
11. If resuspension proves difficult, dislodge the pellet and allow resuspension to occur overnight at 4°C.
12. Invert gently several times until the DNA becomes visible.
13. ReadyRed™ can be used instead of chloroform–isoamyl alcohol when removing any traces of phenol from the aqueous phase.
14. Resuspension of the DNA pellet can be achieved either by flicking the side of the microfuge tube, or by storage at 4°C.
15. In the absence of a fluorescence plate reader, or in cases where contaminants may be present in the DNA sample (which can either contribute to or quench the fluorescence), DNA can be quantified by comparison with λ DNA standards in the 50 ng–2 μg range, following electrophoresis using an agarose minigel ***(6)***.
16. When performing border fragment analysis, choose a unique restriction site within the T-DNA. For internal fragment analysis, choose restriction sites either side of the gene of interest, to produce fragments of defined size.
17. Samples have a gelatinous appearance prior to drying. The samples are reduced in the vacuum chamber to a glassy, almost transparent bead. If the samples are not dried sufficiently, the density of the samples is too low to allow them to sink into the wells of the elecrophoresis gel.
18. Digestion can be assessed rapidly at this stage by running a 0.8% (w/v) agarose minigel for 1 h at 75 V. Fully digested plant genomic DNA should be an uninterupted smear of even intensity. If the DNA has not digested completely, re-precipitate the DNA and repeat the restriction enzyme digestion.
19. A continuous voltage must be applied to the gel to prevent the DNA from diffusing from the gel. Capillary blotting must be carried out immediately after gel electrophoresis.
20. Alternatively, use 3 μg of X kb ladder (12 μL of 0.25 μg/μL; Boehringer Mannheim).
21. Mark the underside of one corner of the nylon membrane with a pencil to determine its orientation in relation to the gel.
22. The PCR reaction component volumes ($MgCl_2$, 10× PCR DIG-labeling mixture, *Taq* DNA polymerase, forward and reverse primers), cycling conditions and tem-

plate DNA concentration, usually have to be optimized empirically *(7)* and are dependent on the primers, template, and the thermocycler. To minimize cost, optimization experiments should be performed with unlabeled dNTPs before using the PCR-DIG-labeling mix. The incorporation of DIG molecules can be confirmed by running, on an agarose gel, the PCR product of the labeling reaction next to an unlabeled PCR product. Since the incorporation of DIG-11-dUTP increases the molecular weight of the PCR product, the product from the labeling reaction appears slightly larger than when a reaction is performed using unlabeled dNTPs.

23. Ensure DIG Easy Hyb solution is at 37°C before use.
24. A temperature of 37°C gives the lowest stringency with DIG Easy Hyb solution. If the probe gives problems with background, increase the hybridization temperature to 42°C. If the problem persists, consult **ref.** ***(3)***.
25. When reusing stored hybridization solution, denature the solution by heating to 68°C for 10 min, followed by cooling on ice for at least 10 min.
26. Ensure the temperature of wash solution B is 65°C before use. The hybridization oven can be allowed to reach 65°C gradually once wash solution B is added. If the probe binds nonspecifically, the temperature of the posthybridization wash can be increased to 68°C and the washing stringency can be increased by substituting 0.1× SSC for 0.5× SSC in wash solution B.
27. A precipitate will result in high background during chemiluminescent detection.
28. Ensure anti-DIG-AP solution and buffer C are mixed thoroughly before use.
29. Do not allow SDS to boil.

References

1. McCabe, M. S., Power, J. B., de Laat, A. M. M., and Davey, M. R. (1997) Detection of single-copy genes in DNA from transgenic plants by nonradioactive Southern blot analysis. *Mol. Biotechnol.* **7,** 79–84.
2. Dellaporta, S. L., Wood, J., and Hicks, J. B. (1983) A plant DNA minipreparation: Version II. *Plant Mol. Biol. Report.* **4**, 19–21.
3. Boehringer Mannheim GmbH, Biochemica (1993) *The DIG System User's Guide for Filter Hybridisation*, Mannheim, Germany.
4. Saiki, R. K., Scharf, S., Faloona, F., Mullis, K. B., Horn, G. T., Erlich, A. H., and Arnheim, N. (1985) Enzymatic amplification of β-globin genomic sequences and restriction site analysis for diagnosis of sickle cell anemia. *Science* **230,** 1350–1354.
5. Lion, T. and Haas, O. A. (1990) Nonradioactive labeling of probe with digoxigenin by polymerase chain reaction. *Anal. Biochem.* **188,** 335–337.
6. Sambrook, J., Fritsch, E. F., and Maniatis, T., eds. (1989) *Molecular Cloning: A Laboratory Manual*, Vol. 3, 2nd edit., Cold Spring Harbor Laboratory Press, Cold Spring Harbor, New York.
7. Cobb, B. D. and Clarkson, J. M. (1994) A simple procedure for optimising the polymerase chain reaction (PCR) using modified Taguchi methods. *Nucleic Acids Res.* **22,** 3801–3805.

21

Differentiation Between Transcripts of Genes Belonging to Small Families During Fruit Ripening and Abscission

Claudio Bonghi, Benedetto Ruperti, and Pietro Tonutti

1. Introduction

Since its development in 1983 *(1)*, Ribonuclease Protection Assay (RPA) has become a widely employed technique for RNA analysis. The procedure is based on hybridization of the RNA being analyzed to a labeled (nonisotopic or radioactive) antisense RNA probe with successive digestion of unhybridized RNA with a cocktail of single-strand-specific RNases (usually RNase A and T1). Only hybridized probe is protected from digestion and, after separation on a polyacrylamide gel, can be visualized on a film. As long as the probe is present in the hybridization solution in molar excess over the target message, the signal intensity is proportional to the starting amount of complementary transcript, thus enabling a quantitative analysis of gene transcription. Compared to hybridization techniques relying on the use of targets bound to a solid support (i.e., Northern analysis), RPA ensures a considerably higher sensitivity together with consequently reduced exposure time and background. Rare messages can be detected and quantitative comparisons can be performed using an internal control probe for RNA loading normalization. Furthermore, specificity of the target sequence recognition and therefore of the final signal is enhanced by the RNase digestion step. Small differences between the probe and the complementary mRNA can be in fact detected by adjusting the ribonuclease concentration so that only perfect matches will be protected. This makes RPA particularly suitable for transcription analysis of gene family members. One possible limitation of RPA is that the conditions of digestion and hybridization must be optimized to obtain reproducible results, making the procedure

From: *Methods in Molecular Biology, vol. 179: Gene Probes: Principles and Protocols*
Edited by: M. Aquino de Muro and R. Rapley © Humana Press Inc., Totowa, NJ

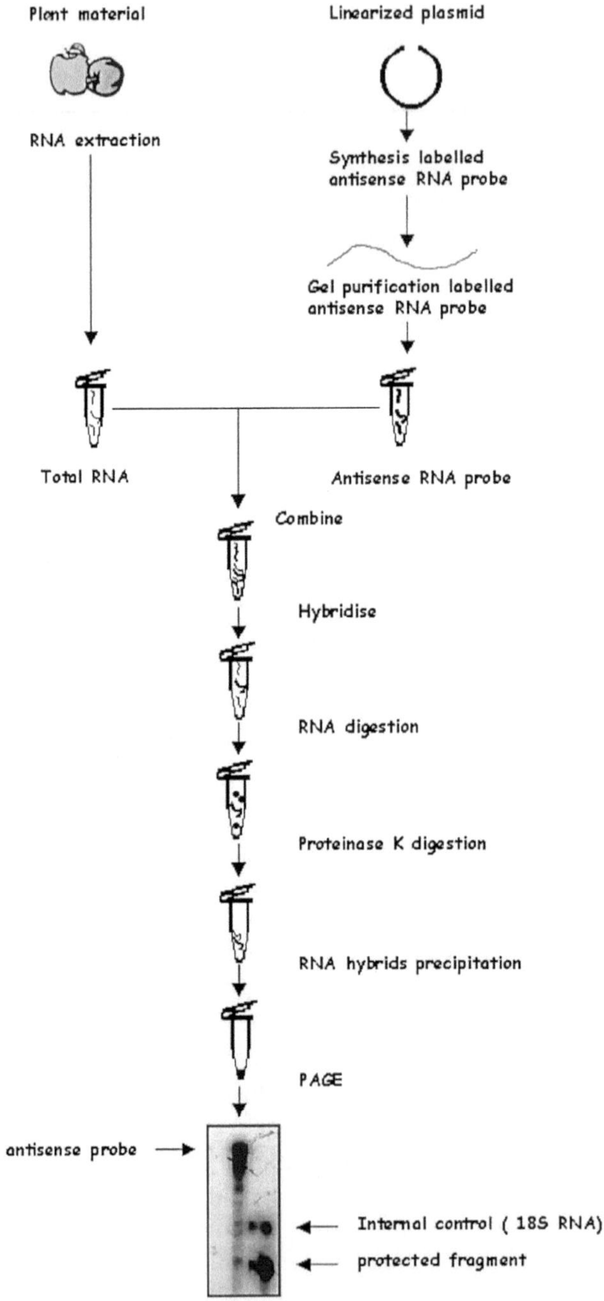

Fig. 1 A Diagramatic summary of the RPA method

somehow time consuming. However, commercially available kits and a number of improvements have been introduced to make the procedure more straightforward. RPA has been successfully used to discriminate with a high degree of specificity and sensitivity between transcripts of genes belonging to highly conserved plant multigene families by using their 3' untranslated regions as probes *(2–5)*. The following protocol (a schematic representation is shown in **Fig. 1**) has proved in our experience to be a good compromise in terms of quality of results and ease of use, and relies on isolation of good quality RNA as a prerequisite. We have successfully used RPA in detecting low-level transcripts of cell wall hydrolases *(6)* and of ethylene biosynthetic genes *(7)* involved in fruit ripening and abscission. This makes RPA a powerful tool to study plant gene transcription when specific detection of low-level or of highly homologous transcripts is a primary need.

2. Materials

2.1. Preparation of RNA Probe

2.1.1. Isolation and Cloning of DNA Fragment

1. Restriction endonucleases.
2. Agarose gel electrophoresis apparatus and electrophoresis grade agarose. For optimal resolution of DNA fragments <500 bp, NuSieve GTG agarose (FMC Bioproducts, Rockland, ME) is recommended.
3. DNA size marker (*Hae*III-digested ϕX-174, Promega, Madison, WI).
4. Agarose DNA purification kit QIAquick (Qiagen,Valencia, CA).
5. Plasmid carrying T7 and T3 or SP6 promoters (pGEM series by Promega).
6. TE: 10 m*M* Tris-HCl, pH 8.0; 1 m*M* EDTA.
7. T4 DNA ligase and ligation buffer (Promega).
8. TOP10F' competent cells (Invitrogen, Carlsbad, CA).
9. LB, 2× YT, SOC media, LB plates (LB medium and agar), 5-bromo-4-chloro-3-indolyl-β-D-galactoside (X-Gal), isopropyl-β-D-thiogalactoside (IPTG), and antiobiotics.
10. Plasmid Minikit (Qiagen).
11. Proteinase K (Promega).
12. DNA sequencing kit (PE Applied Biosystems, Foster City, CA).

2.1.2. Synthesis of Labeled RNA Probe

1. Diethylpyrocarbonate (DEPC)-treated water.
2. 5× Transcription buffer: 200 m*M* Tris-HCl, pH 8.0, 250 m*M* NaCl, 40 m*M* $MgCl_2$, 10 m*M* spermidine.
3. T7, T3, or SP6 RNA polymerases (Ambion, Austin, TX).
4. Rnasin (Promega).

5. [α-^{32}P]UTP (10 mCi/mL, 400—800 Ci/mmol).
6. DNase I, RNase-free (Promega).

2.1.3. Gel Purification of RNA Probe

1. Polyacrylamide gel electrophoresis apparatus.
2. Acrylamide, *bis*-acrylamide,
3. 1× TBE: 89 m*M* Tris, 89 mM boric acid, 2 m*M* EDTA.
4. TEMED.
5. 10% Ammonium persulfate.
6. Urea.
7. Loading buffer: 95% Formamide, 0.025% xylene cyanol, 0.025% bromophenol blue, 0.5 m*M* EDTA, 0.025% sodium dodecyl sulfate (SDS).
8. X-OMAT AR X-ray film (Eastman Kodak, Rochester, NY).
9. Elution buffer: 0.5 *M* NH_4OAc, 10 m*M* $Mg(OAc)_2$, 1 m*M* EDTA, 0.1% SDS.

2.2. RPA Analysis

2.2.1. RNA Extraction

1. Extraction buffer: 100 m*M* Tris-HCl, pH 9.0; 100 m*M* NaCl, 5 m*M* EDTA, 2.5% β-mercaptoethanol; 1% SDS; 1% polyvinyl pyrrolidone (PVP); 1% polyvinyl polypyrrolidone (PVPP). 5 m*M* ascorbic acid, and 100 µg/mL proteinase K.
2. Resuspension buffer: 25 m*M* boric acid, 50 m*M* Tris-HCl, pH 7.6; 1.25 m*M* EDTA pH 8.0 100 m*M* NaCl.
3. 2-Butoxyethanol.

2.2.2. RNA Sample Preparation and Hybridization

1. Yeast RNA.
2. Hybridization solution: 80% deionized formamide, 40 m*M* PIPES, pH 6.4, 0.4 *M* NaCl, 1 m*M* EDTA.

2.2.3. RNA Digestion

1. Ribonuclease digestion buffer: 10 m*M* Tris-HCl, pH 7.5, 300 m*M* NaCl, 5 m*M* EDTA, pH 8.0.
2. RNase-ONE Ribonuclease (Promega).
3. Proteinase K (Promega).

2.2.4. Separation and Detection of the Protected Fragment

1. Polyacrylamide gel apparatus (*see* **Subheading 2.1.3.**).
2. RNA loading buffer: 95% formamide, 0.025% xylene cyanol, 0.025% bromophenol blue, 0.5 m*M* EDTA, 0.025% SDS.
3. Labeled marker (100 basepair [bp] ladder, Promega).

3. Methods

3.1. RNA Probe Preparation

3.1.1. Isolation and Cloning of DNA Fragments

1. Incubate plasmid vectors (up to 5 µg) containing genomic DNA or cDNA with the appropriate restriction endonucleases (*see* **Note 1**) and buffers in a final volume of 20 µL for 2–3 h to obtain templates sizing from 100 to 300 bp (*see* **Note 2**).
2. Prepare 2% agarose gel containing ethydium bromide (10 µg/mL), and load the digestion products and DNA size marker (*Hae*III-digest of ϕX-174). Select fragments by comparison with the size marker.
3. Cut with a razor blade the gel slice containing the selected fragment and purify with the QIAquick gel extraction kit (Qiagen) (*see* **Note 3**). Quantify the concentration by spectrophotometer. Adjust the concentration with sterile water to 25 ng/µL.
4. Digest plasmid vector (5 µg) bearing the T7 and SP6/T3 promoters with appropriate endonuclease and buffer in a total volume of 20 µL. Verify digestion by running an agarose (1%) gel loaded with 0.5 µg of the digested plasmid.
5. Add 30 µL of sterile water to reaction mixture and extract with equal volume of phenol preequilibrated with TE buffer. Vortex-mix. Add 50 µL of chloroform–isoamyl alcohol (24:1). Vortex-mix. Centrifuge in a microfuge at room temperature for 30 s to separate the phases. Remove the upper phase and place it in a clean tube.
6. Precipitate plasmid DNA by addition of 1/10 volume of 4 *M* sodium acetate, vortex-mix briefly, than add 2.5 volume of cold 95% ethanol. Incubate at –20°C for at least 3 h.
7. Recover vector plasmid by centrifugation (10,000*g*) in a microfuge at 4°C for 30 min. Pour off supernatant and wash 2× with 500 µL of cold 70% ethanol.
8. Resuspend the pellet in sterile water, quantify by spectrophotometer, and adjust the concentration at 25 ng/µL.
9. In a microfuge tube mix 1 µL of vector plasmid, 0.3 µL of DNA template solution, T4 DNA ligase (4 Weiss units), and 1 µL of T4 10× ligation buffer (*see* **Note 4**). Add sterile water to reach total volume of 10 µL. Incubate ligation reaction at 14°C overnight.
10. Thaw on ice 0.5 *M* β-mercaptoethanol (β-ME) and 50 µL of competent cells (*see* **Note 5**).
11. Pipet 2 µL of the 0.5 *M* β-ME into the vial of the competent cells and mix gently with the pipet tip. Do not mix by pipetting up and down.
12. Pipet 1 µL of the ligation reaction into the vial of the competent cells and incubate on ice for 30 min.
13. Heat-shock for exactly 30 s in the 42°C water bath.
14. Remove the vial from the 42°C water bath and place it on ice for 2 min.
15. Add 450 µL of SOC medium at room temperature.
16. Shake the vial horizontally at 37°C for 1 h at 225 rpm in a rotary shaking incubator.

17. Place transformed cells on ice.
18. Spread 50–100 µL of transformed cells on LB agar plates containing appropriate antibiotic(s) with X-Gal (40 µL of a solution of 40 mg/mL) and IPTG (40 µL of a solution of 100 m*M*).
19. Incubate at least for 18 h at 37°C.
20. Pick positive clones (white colonies) and inoculate into tubes containing 2× YT broth and appropriate antibiotic(s).
21. Incubate at 37°C for at least 16 h or until broth is moderately turbid.
22. Make a plasmid DNA miniprep using Qiagen plasmid minikit.
23. Digest plasmid DNA with proteinase K (100–200 µg/mL) for 30 min at 50°C and then extract with phenol–chloroform to remove RNase used during the isolation procedure.
24. To determine the insert orientation, carry out sequencing analysis using universal primers (*see* **Note 6**).

3.1.2. Preparation of the DNA Template

1. Linearize the plasmids (5 µg in a 20-µL total volume) containing the DNA templates and the DNA to be used as internal control (*see* **Note 7**) by restriction endonuclease that cleaves the plasmid distally of the promoters (*see* **Note 8**).
2. Check the complete digestion by running an aliquot on 1% agarose gel.
3. Add 30 µL of sterile water to reaction mixture and extract with 50 µL of phenol preequilibrated with TE buffer. Vortex-mix. Add 50 µL of chloroform–isoamyl alcohol. Vortex-mix. Centrifuge in a microfuge at room temperature for 30 s to separate the phases. Remove the upper phase and place it in a clean tube.
4. Precipitate plasmid template by addition of 1/10 volume of 4 *M* sodium acetate, vortex-mix briefly, than add 2.5 volume of cold 95% ethanol. Incubate at –20°C for at least 3 h.
5. Recover DNA plasmid by centrifugation in a microfuge at 4°C for 30 min. Pour off supernatant and wash 2× with 500 µL of cold 70% ethanol.
6. Redissolve at 1 µg/µL in RNase-free TE buffer .

3.1.3. Synthesis of Labeled RNA Probe

1. Place the required tubes for the labeling reaction, excluding the enzyme, on ice to thaw.
2. Mix in a 1.5-mL microfuge tube at room temperature (*see* **Note 9**) the labeling components at the following order:
   4 µL of 5× transcription buffer
   2 µL of 0.1 *M* dithiothreitol (DTT)
   0.8 µL of placental ribonuclease inhibitor (25 U/µL)
   4 µL of 2.5 m*M* NTP mix (A, C, G)
   2.4 µL of 100 µ*M* cold UTP (*see* **Note 10**)
   1 µL of 500 ng/µL of linearized DNA template or a same amount of internal control
   5 µL of [α-^{32}P] UTP (10 µCi/µL)

1 µL of bacteriophage RNA polymerase (5–10 U) SP6, T7, or T3.

3. Incubate for a minimum of 1 h at 37°C for T7 and T3, at 40°C for SP6.
4. Transfer microfuge tube at 95°C for 2 min.
5. Add 1 µL RNase-free DNase I (2 U/µL) and incubate at 37°C for 15 min.

3.1.4. Gel Purification of RNA Probe

1. At the end of DNase I incubation, add an equal volume of gel loading buffer, and heat the tube for 3–5 min at 85–95°C.
2. Remove 2 µL for the calculation of yield and specific activity of the probe (*see* **Note 11**).
3. Load the reaction on the 0.75–1 mm thick 5% polyacrylamide 8 *M* urea small sequencing-type gel (PAGE) (*see* **Note 12**) and run until the bromophenol blue approaches the bottom of the gel (about 20 min to 1 h at 100–300 V).
4. After electrophoresis, remove one glass plate and cover the gel with plastic wrap and expose to X-ray film for 2–3 min (*see* **Note 13**).
5. Cut the corresponding labeled band on the gel and transfer with sterile forceps to microfuge tube and submerge with 300 µL of elution buffer.
6. Incubate the tube at 37°C for at least 10 h (*see* **Note 14**).
7. Centrifuge at 3000 rpm for 2 min and recover the supernatant.
8. Reextract gel slice with 100 µL of elution buffer. After centrifugation, recover and combine the supernatants.
9. Extract once with equal volume of phenol–chloroform and add 1/10 of 3 *M* NaOAc, pH 6, and 2.5 volume of 100% ethanol.
10. Incubate at –70°C for 1 h.
11. After precipitation (10,000*g* for 30 min at 4°C) pour off the supernatant, wash the pellet with ethanol (95%), dry under vacuum for few minutes, and redissolve in small volume (about 20 µL) of DEPC-water.
12. Determine specific activity of the eluted probe by scintillation counting.
13. Store the probe at –20°C.

3.2. RPA Analysis

3.2.1. RNA Extraction

All steps must be carried out in an RNase-free environment.

1. Grind 4 g of frozen fruit tissue to a fine powder in a prechilled mortar and pestle with liquid nitrogen.
2. Using a metal spatula quickly transfer the pulverized tissue to a 38-mL polypropylene tube containing 10 mL of preheated (at 65°C) extraction buffer.
3. Incubate for 10 min at 65°C.
4. Centrifuge in swing-out rotor at 20,000*g* for 5 min to eliminate the debris.
5. Recover the supernatant and add equal volume of TE-equilibrated phenol warmed at 65°C.
6. Mix and centrifuge as indicated previously.

7. Transfer supernatant to a polypropylene tube and add equal volume of phenol–chloroform–isoamyl alcohol (25:24:1).
8. Shake sample and centrifuge as described in **step 4**.
9. Repeat the supernatant extraction with chloroform–isoamyl alcohol.
10. Recover aqueosus phase in a Corex tube and add 1/10 volume 3 *M* NaOAc, pH 4.8, and 2.5 volumes of EtOH to precipitate nucleic acids. Store for 1 h at –80°C.
11. Centrifuge at 12,000*g* for 25 min at 4°C and discard the supernatant.
12. Wash the pellet with 70% ethanol 2×. Dry the pellet for 2–3 min, maintaining the Corex tube upside down.
13. Resuspend in 10 mL of resuspension buffer.
14. Add 4 mL of butoxy ethanol (2-BE), vortex-mix, and store in ice for 30 min (*see* **Note 15**).
15. Centrifuge for 10 min at 4°C at 12,000*g* and recover the supernatant.
16. Add 6 mL of 2-BE, vortex-mix, and keep on ice for 30 min.
17. Centrifuge as indicated in **step 15** and save the pellet.
18. Wash the pellet with cold 70% ethanol, and centrifuge at 4°C at 12,000*g* for 5 min (*see* **Note 16**).
19. Redissolve the pellet in 6 mL of 1× TBE and add 2 mL of 8 *M* LiCl to reach a final concentration of 2 *M* LiCl.
20. Seal the tube with parafilm and mix the content well by inversion. Precipitate overnight on ice at 4°C.
21. Pellet the RNA by centrifugation at 12,000*g* at 4°C for 30 min. Carefully discard the supernatant and rinse the pellet with cold 70% ethanol, and centrifuge at 12,000*g* at 4°C and pour off the supernatant. Repeat.
22. Resuspend the pellet in 200 µL of DEPC-water and transfer to a microfuge tube.
23. To assess the quality and the concentration of the RNA, determine the A_{230}, A_{260}, A_{280}, A_{320} of a 1:99 dilution in sterile water (*see* **Note 17**).

3.2.2. RNA Sample Preparation and Hybridization

1. In a vacuum evaporator centrifuge dry 40 µg of RNA sample (*see* **Note 18**) and, in two additional microfuge tubes, 10 µg of yeast RNA as control (*see* **Note 19**).
2. Resuspend samples in 30 µL of hybridization solution by vortex-mixing.
3. Add diluted ^{32}P-UTP-labeled probe (5×10^5 cpm), heat for 2–3 min at 95°C, and then hybridize at 45°C overnight (*see* **Note 20**).

3.2.3. RNase Digestion

1. Add to the RNA sample and yeast RNA tubes 350 µL of ribonuclease digestion buffer. One microliter of RNase ONE is added to the RNA sample and one of the yeast RNA. Incubate at 30°C for 45 min.
2. Add 10 µL of 20% SDS and 2.5 µL of 20 mg/mL proteinase K. Incubate for 15 min at 37°C.
3. Extract once with 400 µL of phenol–chloroform–isoamyl alcohol (50:48:2). Remove the aqueous phase to a clean microfuge tube containing 1 µL of 10 mg/mL yeast tRNA. Add 1 mL of absolute ethanol and incubate at –20°C for 15–30 min.

4. Centrifuge at 4°C at 10,000*g* for 30–60 min. Pour off the supernatant and dry the pellet in air.

3.2.4. Separation and Detection of the Protected Fragments

1. Prepare a 5% polyacrylamide 8 *M* urea gel at least 11 cm long measured from the bottom of the wells.
2. Resuspend each sample pellet in RNA loading buffer (4–8 µL). The yeast RNA sample without RNase should be resuspended in 40 µL of RNA loading buffer.
3. Dilute a sample of the probe and a molecular weight standard (100-bp ladder end-labeled by Klenow with dCTP) to 1000 cpm in 8 µL of loading buffer (*see* **Note 21**).
4. Heat all the samples at 95°C for 3–5 min and quench on ice.
5. Rinse the wells of the gel and immediately load the entire sample volumes (except for the yeast RNA control sample without RNase, where only 4 µL are loaded).
6. Run the gel until the bromophenol blue runs off the gel.
7. Dry the gel and expose to X-ray film with an intensifying screen.

4. Notes

1. Restriction enzymes must be chosen to isolate fragments corresponding to coding or 3' untranslated regions. If such enzymes are not available, specific primers for PCR amplification should be selected ***(3)***.
2. DNA templates exceeding 300 nucleotides in length may result in incomplete transcripts due to ribonuclease contamination or termination of the RNA polymerase before completion of the transcript. These phenomena induce high background in the hybridizations.
3. Alternative methods can be used for the purification of DNA from agarose: freeze-squeeze method ***(9)***, Centricon-100 spin filtration column (Amicon, Beverly, MA), GeneClean method (Bio 101, La Jolla, CA).
4. The vector plasmid: DNA template ratio should be in the range of 1:1 to 1:3. To estimate the amount of DNA template needed to ligate with 25 ng of vector at the ratio of 1:1, the following formula can be used: *x* ng of DNA template = (*y* bp of DNA template) (25 ng vector)/ size in bp of the vector.
5. Different competent cells may be selected considering the genotype. We have successfully used TOP10F' that does express the *lac* repressor (*lac*I^q) which will repress transcription from the *lac* promoter. For blue-white colony selection, IPTG must be added to the plates.
6. It is important to determine the template orientation to synthesize antisense and sense RNA probes.
7. Commonly used internal controls (β actin, 18S and 28S rRNAs) are typically constitutively expressed genes.
8. All types of restriction enzymes can be used, although in the literature a low level of transcription has been reported when 3' overhanging ends were produced by

specific enzymes (*Kpn*I, *Pst*I, etc.) ***(10)***. If it is necessary to use a restriction enzyme that cleaves a 3' overhang, a reaction with Klenow should be used to blunt the ends.

9. If the reaction is assembled on ice coprecipitation of DNA template and spermidine in the 5× transcription buffer occurs.
10. The greater the concentration of unlabeled UTP (limiting nucleotide) the lower the specific activity of the transcript. However, the proportion of full-length transcripts increases and the proportion of prematurely terminated transcripts decreases with increasing the unlabeled UTP concentration. The concentration of limiting nucleotide should be ≥3 μ*M* for synthesis of RNA probes up to 400 bp. To synthesize very high specific activity probes omit the unlabeled UTP but in this case the proportion of prematurely terminated transcripts is very high. A compromise to obtain full-length probes with high specific activity is the concentration of 1 μ*M* of labeled UTP and 2.5–10 μ*M* of unlabeled UTP. In some cases, maximizing sensitivity is desirable. For detection of mRNA of unknown abundance no unlabeled limiting nucleotide is used to make a probe of maximum specific activity. Considering its abundance, the internal control probe needs to be synthesized to lower specific activity. To decrease specific activity, increase the amount of unlabeled UTP up to a ratio of labeled to unlabeled UTP of 1:10,000 if 18S or 28S rRNAs are used. This allows to obtain a single exposure in which both the internal control and rare message signals are in the linear range of the film.
11. The efficiency of the transcription reaction can be determined by trichloroacetic acid (TCA) precipitation of the reaction products ***(8)***.
12. Nondenaturing gels are used in the interest of speed, but it is important that the RNA be fully denatured before loading.
13. Exposure time should allow to yield a light band so that a thin gel fragment can be excised from the gel.
14. Approximately 20% of the labeled probe is obtained after 2 h of incubation. To elute about 95% of the probe, incubate overnight.
15. Because fruit tissues contain large amount of carbohydrates, this differential precipitation is necessary to separate carbohydrates from RNA.
16. After washing, the pellet must appear white and not jellylike.
17. Compared to Northern analysis, RPA is relatively insensitive to mRNA degradation, particularly if the probe spans a short (≤ 300 nt) subregion of the mRNA.
18. For most messages, 10 μg or less of total RNA are usually sufficient. For rare messages we have used up to 80 μg ***(6)***.
19. Yeast RNA is used as a control for nonspecific hybridization and completion of the ribonuclease digestion: these hybridization reactions should yield no protected fragment of the predicted size.
20. A good temperature to try is 45°C, but it is advisable to test a range of temperatures of 30–60°C, considering that secondary structures can form within the probe.

21. In denaturing polyacrylamide–urea gels, RNA has a 5–10% lower mobility than DNA of the same size. Thus, if an RNA species runs with a DNA marker of 100 bp its length is 90–95 bp.

References

1. Zinn, K., DiMaio, D., and Maniatis, T. (1983) Identification of two distinct regulatory regions adjacent to the human β-interferon gene. *Cell* **34,** 865–879.
2. Yip, W-K., Moore, T., and Yang, S. F. (1992) Differential accumulation of transcripts for four tomato 1-aminocyclopropane-1-carboxylate synthase homologs under various conditions. *Proc. Natl. Acad. Sci. USA* **89,** 2475–2479.
3. Barry, C. S., Blume, B., Bouzayen, M., Cooper, W., Hamilton, A. J., and Grierson, D. (1996) Differential expression of 1-aminocyclopropane-1-carboxylic acid oxidase gene family of tomato. *Plant J.* **9,** 525–535.
4. Kalaitzis, P., Solomos, T., and Tucker, M. L. (1997) Three different polygalacturonases are expressed in tomato leaf and flower abscission, each with a different temporal expression pattern. *Plant Physiol.* **113,** 1303–1308.
5. Whittaker, D. J., Smith, G. S., and Gardner, R. C. (1997) Expression of ethylene biosynthetic genes in *Actinidia chinensis* fruit. *Plant Mol. Biol.* **34,** 45–55.
6. Tonutti, P., Cass, L., and Christoffersen, R. C. (1995) The expression of cellulase gene family members during induced avocado fruit abscission and ripening. *Plant Cell Environ.* **18,** 709–713.
7. Ruperti, B., Bonghi, C., Tonutti, P., and Ramina, A. (1998) Ethylene biosynthesis in peach fruit abscission. *Plant Cell Environ.* **21**, 731–737.
8. Sambrook, J., Fritsch, E. F., and Maniatis, T. (1989) *Molecular Cloning: A Laboratory Manual*, 2nd edit., Cold Spring Harbor Laboratory, Cold Spring Harbor, NY.
9. Stoflet, E. S., Koeberl, D. D., Sarkar, G., and Sommer, S. S. (1988) Genomic amplification with transcript sequencing. *Science* **239,** 491–494.
10. Schendorn, E. T. and Mierindorf, R. C. (1985) A novel transcription property of SP6 and T7 RNA polymerases: dependence on template structure. *Nucleic Acids Res.* **13,** 6223–6236.

22

Polymerase Chain Reaction Detection of Invasive *Shigella* and *Salmonella enterica* in Food

Keith A. Lampel and Palmer A. Orlandi

1. Introduction

The ability to detect *Shigella* in foods is often hampered by the often-low numbers of organisms present in the sample at the time of analysis. At present, a reliable, time-efficient, and highly sensitive protocol to isolate *Shigella* from foods is unavailable although a conventional method is currently described in the *Bacteriological Analytical Manual* (BAM; *1*). Through the use of a polymerase chain reaction (PCR)-based assay, many of the problems associated with analyzing food samples for the detection of pathogens can be overcome. The preparation of template DNA from pathogenic bacterial cells in sufficient numbers and free of potential PCR inhibitors is quite difficult, as with any given pathogen in a complex matrix.

In the present study, PCR primers that target the *ipaH* gene are used to detect all four species of *Shigella*: *S. dysenteriae* (serogroup A), *S. flexneri* (B), *S. boydii* (C), and *S. sonnei* (D). This gene is present on the large virulence plasmid and also in the chromosome *(2)*. Loss of the virulence plasmid renders *Shigella* avirulent. Whereas the effects of prolonged storage in foods or the environment may result in the loss of the virulence plasmid, a PCR-based detection method using primers directed to the *ipaH* gene will not be affected. Template DNA can effectively be prepared from food washings. Differential centrifugation runs to remove food matter and to concentrate cells are then followed by steps to increase cell numbers by cultural enrichment or to directly lyse isolated bacterial cells. These lysates are suitable material as DNA template for PCR.

From: *Methods in Molecular Biology, vol. 179: Gene Probes: Principles and Protocols*
Edited by: M. Aquino de Muro and R. Rapley © Humana Press Inc., Totowa, NJ

As an alternative PCR template preparation protocol, food washes can be directly applied to special filters designed to eliminate many of the inherent problems associated with such diverse elements as are found in foods. These filters rely on their ability to trap and lyse bacterial cells within the filter; this allows for greater numbers of pathogens to be sequestered in the filter and the convenience of lysing the cells without the need for extraction and purification steps. Brief washes are then necessary to remove particulate debris and potential inhibitors of PCR.

Salmonella are much more ubiquitous and more commonly found in foods than *Shigella*. For detecting the genus *Salmonella*, an oligonucleotide directed to the *invA* gene *(3)* is suggested as a DNA probe in colony hybridization blots *(4)*. This probe is able to detect most of the more than 2000 serovars of *Salmonella*. In some instances, however, the detection of only one serovar in particular is required. In such cases, DNA probes can be synthesized with only one basepair difference between the targeted organism, for example, Enteritidis *(5)*, and all the other salmonellae.

Using PCR to detect the presence of salmonellae in foods, primers were designed to amplify the *invA* gene *(6)*. For more selective amplification, such as for Enteritidis, PCR primers are selected to generate a 351-basepair (bp) DNA fragment from the *Salmonella* plasmid virulence (*spvA*) gene *(7)* of Enteritidis. For sensitive, serovar selectivity, we used Mismatch amplification mutation assay (MAMA, *8*), a PCR protocol that uses primers with a single base mismatch with the wild-type strain. This mismatch allows for the discrimination of single-basepair differences by PCR. A single base difference at position at 271 is present between the nucleotide sequence of the *spvA* gene of Enteritidis and other salmonellae *(9)*. To develop a PCR assay specific for Enteritidis, the downstream 21 base primer was designed to have a single base change incorporated at the penultimate position, nucleotide 20, corresponding to position 270 in the *spvA* gene. This resulted in a single base mismatch with Enteritidis and a two-base mismatch with other salmonellae in the *spvA* gene. The upstream primer is homologous to the *spvA* gene for all *Salmonella*. The annealing step in the thermocycling program is then conducted at a temperature that adds an additional level of stringency to discriminate between strains.

2. Materials

2.1. Food Sampling

1. Phosphate-buffered saline (PBS), pH 7.4.
2. Poly-prep chromatography columns (Bio-Rad, Hercules, CA).
3. Glass wool (Corning, cat. no. 3950).
4. Tris-EDTA (TE) buffer, pH 8.0: 10 m*M* Tris-HCl, 1.0 m*M* EDTA, pH 8.0.

2.2. PCR

1. Thin-walled PCR tubes.
2. Heat-stable DNA polymerase (10× buffer supplied by manufacturer).
3. 10 m*M* dNTP (nucleotide) mix.
4. PCR primers:
   ipaH
      ipaHF: 5'-GTTCCTTGACCGCCTTTCCGATACCGTC-3'
      ipaHR: 5'-GCCGGTCAGCCACCCTCTGAGAGTAC-3'
   ipaH nested
      ipaHNF: 5'-CCACTGAGAGCTGTGAGG-3'
      paHNR: 5'-TGTCACTCCCGACACGCC-3'
   invA
      invAF: 5'-TATCGCCACGTTCGGGCAA-3'
      invAR: 5'TCGCACCGTCAAAGGAACC-3'
   spvA
      mamaF: 5'-GCAGACATTATCAGTCTTCAGG-3'
      mamaR: 5'-TCAGGTTCGTGCCATTGTCAA-3'
5. FTA filters (Fitzco, Maple Plain, MN).
6. FTA purification buffer (GIBCO/BRL, Grand Island, NY).
7. Nonfat dry milk, 10% (w/v) in water.
8. Mineral oil.
9. Thermocycler.

2.3. Gel Electrophoresis

1. Agarose.
2. 10 mg/mL of Ethidium bromide (GIBCO/BRL, Grand Island, NY); **handle with care—teratogenic**.
3. Molecular weight ladder (preferably 100-bp ladder).
4. 10× Tris-acetate-EDTA (TAE) buffer: 400 m*M* Tris-acetate, 10 m*M* EDTA, pH 8.3.
5. 6× Loading dye: 0.25% [w/v] bromophenol blue, 0.25% xylene cyanol FF [w/v], 40% [w/v] sucrose in water.
6. Equipment gel electrophoresis chamber; power supply.

2.4. Hybridization Techniques

1. Nonradioactive hybridizations buffers:
   Buffer 1: 100 m*M* Tris-HCl, 150 m*M* NaCl, pH 7.5
   Buffer 2: Blocking reagent-1% (w/v) of nonfat dry milk in buffer 1.
   Buffer 3: 100 m*M* Tris-HCl, 100 m*M* NaCl, 50 m*M* $MgCl_2$, pH 9.5.

3. Methods

3.1. Food Washing

Food samples, placed in 250-mL sterile beakers, are washed once with 10 mL of 1× PBS for several minutes. The wash buffer is decanted into polyprep chromatography columns packed with glass wool to remove large particulates. Columns can be washed once with an additional 3 mL of 1× PBS. The filtrate was collected in 12.5-mL polypropylene tubes, centrifuged at 8000*g* for 5 min, the supernatant discarded, and the cell pellet suspended in 100 µL of 1× PBS. Ten microliters of the suspended pellet was either transferred to PCR tubes or applied to FTA filters and processed as described in **Subheading 3.2**.

3.2. PCR Detection

3.2.1. Conventional DNA Template Preparation from Bacterial Colonies (Gram-Negative)

Add a fraction of a colony to 150 µL of distilled water contained in a 1.5-mL tube and boil for 5 min. Place tube on ice and use immediately or store at –20°C. Amounts to use are indicated for each PCR setup.

3.2.2. Conventional DNA Template Preparation from Dilutions or Washes of Food Samples

Place 10 µL into a thin-walled PCR tube, add a few drops of mineral oil, and boil for 5 min. Cool the tubes on ice for 1 min before adding the PCR reagents.

3.2.3. Enrichment for PCR

Twenty-five grams of food sample are added to *Shigella* broth (composition per liter: 20 g of pancreatic digest of casein, 5 g of NaCl, 2 g of K_2HPO_4, 2 g of KH_2PO_4, 1 g of glucose, 1.5 mL of Tween-80, pH 7.0) supplemented with 0.5 µg/µL of novobiocin; for *Salmonella*, use lactose broth without any antibiotic. The culture is grown for 20 h under anaerobic conditions at 43°C; for *Salmonella*, in a shaking water bath at 37°C overnight. One milliliter of culture is removed, centrifuged at 320*g* for 3 min to remove large particles and the supernatant transferred to a clean 1.5-mL centrifuge tube. The bacterial cells are pelleted at 8000*g* for 5 min and the supernatant aspirated. Cells are suspended in 1 mL of 1× PBS, vortex-mixed, and centrifuged at 8000*g* for 5 min. Next 100–200 µL of 1× PBS is added, and the cells suspended and boiled for 5 min. For PCR analysis, 2.5 µL of lysate is used.

3.2.4. FTA Filter-Based DNA Template Preparation

Ten microliters of sample (bacterial lysates, dilutions, washes) are applied to FTA filters. After the application of bacterial cells, the filters are air-dried on a heating block at 56°C (15–20 min). The FTA filters were washed twice with 0.5 mL of FTA purification buffer for 2 min, and washed twice in TE buffer for 2 min. Filters were air-dried as described above and either stored at –20°C or the spotted area removed with a 6-mm diameter hole puncher and directly used as template for PCR (*see* **Notes 1** and **2**).

3.2.5. Shigella Detection by PCR

3.2.5.1. Conventional PCR protocol (*see* **Note 3**):

Buffer (10×)	2.5 µL
dNTP (10 m*M*)	0.3 µL
Primers (10 pmol/µL)	2.5 µL each
Template	1–5 µL
Enzyme (1.5 U/25 mL reaction)	0.3 µL
Total volume is brought to 25 µL with distilled water.	
Mineral oil	two drops

3.2.5.2. FTA filter-Based PCR Protocol

Buffer (10×)	10 µL
dNTP (10 m*M*)	1 µL
Primers (10 pmol/µL)	10 µL each
Template	Filter
10% (w/v) nonfat powdered milk	2 µL
Enzyme (2.5 U/100 µL reaction)	1 µL
Total volume is brought to 100 µL with distilled water.	
Mineral oil	four drops

3.2.5.3. Primers

Oligonucleotides were synthesized to target the *ipaH* gene of *Shigella*. If nested PCR is desired for further confirmation, then those primers directed toward internal sites of the first set of PCR primers were used in a secondary reaction.

3.2.5.4. Nested PCR

One to five microliters of the primary PCR product is used as template in a 25-µL reaction volume.

3.2.5.5. Thermocycler Parameters

An initial 5-min denaturation step at 95°C was used. The temperature was then reduced and held at 80°C during the addition of *Taq* DNA polymerase. A 30-cycle program was run in which each cycle had the following steps: denaturation: 1 min at 94°C; annealing: 1 min at 60°C; extension: 1 min at 72°C. An additional step is set for 7 min at 72°C with a final step set at 4°C (*see* **Note 4**).

3.2.5.6. Amplicons

The first set of primers yields a PCR product of 620 bp whereas the nested PCR product is 295 bp.

3.2.6. *Salmonella* Detection by PCR

Buffer (10×)	2.5 μL
dNTP (10 m*M*)	0.3 μL
Primers (10 pmol/μL)	2.5 μL each
Template	1-5 μL
Enzyme	0.3 μL
Total volume is brought to 25 μL with distilled water.	
Mineral oil	two drops

3.2.6.1. Thermocycler Parameters

For three-step amplification, for example, using the invA primers, denaturation is at 94°C for 1.0 min, annealing of primers at 60°C for 1.0 min, and extension at 72°C for 1.0 min. In two-step amplification, the denaturation step is at 94°C for 1.0 min and the annealing and extension steps are combined into one step at 64°C for 1.5 min.

3.2.6.2. Amplicons

The invA primers yield a 275-bp PCR product; the SE-MAMA primers yield a 351-bp product.

3.2.7. Electrophoresis of PCR Products

PCR-amplified products (10–15 μL with 2–3 μL of dye) are visualized in 1% agarose gels in 0.5× TAE buffer, pH 7.8, with ethidium bromide. In a small electrophoresis apparatus, such as a Run-One gel system (Embi-Tec, San Diego, CA), run at 100 V. Larger gels can be run at higher volts, between 125 and 150 V. A 100-bp ladder is used as molecular weight marker.

3.3. *Salmonella*-Gene Probe

Colony hybridization is a convenient way to analyze food samples for the presence of pathogenic bacteria. Food samples can be initially added to enrichment broth, for example, lactose broth, to increase the number of target organisms, and then after overnight growth, plated onto differential and selective agar media for *Salmonella*, such as Hektoen enteric, xylose–lysine–desoxycholate agars. In some instances, colonies from agar plates are inoculated into broth in a 96-well cluster plate and grown overnight at 37°C. Isolates are replicated onto tryptone soya agar plates and grown at 37°C overnight.

3.3.1. Filter Preparation

3.3.1.1. Radioactive Hybridizations

1. Bacterial colonies are transferred to Whatman 541 filter paper, placed in a glass Petri dish colony side up containing 5 mL of 0.5 *M* NaOH plus 1.5 *M* NaCl, and exposed microwave irradiation for 30 s at 30% power setting (750 W) (*Note*: if the filters tend to rise to the top of the Petri dish, due to the high number of cells or food matrix on the filter, 2 min 15 s irradiation at 30% power can be used).
2. This is followed by neutralization in another Petri dish containing a Whatman 3 filter soaked in 5 mL of 1.0 *M* Tris, pH 7.0, plus 2.0 *M* NaCl for 5 min. Filters are then air-dried on absorbent paper and stored until needed.

3.3.1.2. Nonradioactive hybridizations:

1. Overnight plate cultures were transferred to magnagraph nylon membranes according to the Genius Systems User's Guide (Boehringer Mannheim Biochemicals, Mannheim, Germany). Culture plates are chilled for 1 h at 4°C. Membranes are placed on the plates for 1 min and then transferred, colony side up, onto Whatman No. 3 filter paper.
2. Colonies on filter are denatured in 0.5 *M* NaOH plus 1.5 *M* NaCl for 5 min and neutralized on filter paper with 1.0 M Tris, pH 7.0, plus 2.0 *M* NaCl for 5 min.
3. Membranes are baked for 2 h (30 min minimum) in a vacuum oven at 80°C. Alternatively, crosslinking with UV light is acceptable.

3.3.2. Oligonucleotide Probes

In some cases, probes can be designed to detect one serovar, such as Enteritidis (5'-GCAGACAC**T**GGACAATGG-3'; position no. 9 in the probe, in bold, is the sequence from Enteritidis; other salmonellae differ in sequence at this position) ***(5)***. Oligonucleotides can be designed with one basepair difference between one serovar and another (or many) and discriminate between the selected target and all other salmonellae. An example is a probe directed to the *spvA* gene of Enteritidis that is specific only for this serovar; the one basepair difference under stringent hybridization and wash conditions described are sufficient to distinguish between Enteritidis and other salmonellae.

3.3.2.1. Radioactive Labeling

Probes are labeled using [γ-^{32}P]ATP (spec act of 3000 Ci/m*M*) and T4 polynucleotide kinase and purified over a NucTrap column (or other suitable column).

3.3.2.2. Nonradioactive Labeling

Probes are labeled with digoxygenin (DIG)-dUTP using the DIG oligonucleotide tailing kit from Boehringer Mannheim.

3.3.3. Hybridization

The protocol is given for the oligonucleotide probe to detect Enteritidis; 56°C is used for hybridization and washes.

3.3.3.1. Hybridizations with Radioactive Probes

1. Filters are prehybridized at the theoretical dissociation temperature (T_D) in separate Petri dishes containing 8–10 mL of hybridization mixture: 5× Denhardt solution, 6× SSC, 1 m*M* EDTA, and 0.2 mg of sonicated calf thymus DNA per milliliter (1 mL of the sonicated calf thymus DNA is heated in a boiling water bath for 5 min just prior to addition to the hybridization mix).
2. After 1 h, filters are placed in fresh hybridization mixture containing approximately 3–5 × 10^6 cpm of labeled probe, incubated from 2-3 hours to overnight at the T_D with gentle shaking at 75 rpm (*see* **Note 5**).
3. Filters are washed three times (30 min each) at the T_D with prewarmed 6× SSC containing 0.1% SDS. Rinse briefly in 1X SSC at room temperature, drain filter on absorbent paper but do not allow filters to dry, cover filter in plastic wrap.
4. Place filter(s) in cassette, lay X-ray film on top of filter (between filter and intensifying screen) and store at –70·C for 4–16 h.

3.3.3.2. Hybridizations with Nonradioactive Probes (*see* Note 6)

1. Filters are hybridized with 5 pmol of labeled probe (1–10 pmol is used for end-labeled probes) for 4 h at the T_D. The hybridization solution with the labeled probe is removed (the solution can be stored at –20°C and can be reused several times).
2. Filters are washed twice for 5 min with 2× SSC with 0.1% SDS at room temperature followed by two washes with 0.1× SSC with 0.1% SDS at T_D for 15 min each.
3. The last wash solution is removed and filters are blocked with buffer 2 for 30 min.
4. Immunological detection was performed using a nucleic acid detection kit (Boehringer Mannheim). Dilute alkaline phosphatase (AP)-conjugated anti–digoxygenin Fab fragments (antibody-conjugate) to 150 mU/mL in buffer 2.

5. Remove buffer 2 from filters and briefly wash in buffer 1. Filters are incubated with 20 mL of diluted antibody–conjugate solution for 30 min at room temperature.
6. Filters are washed for 30 min in buffer 1 to remove unbound antibody–conjugate, then equilibrated for 2 min in 20 mL of buffer 3, and color developed in 10 mL of freshly prepared AP substrates (45 µL of nitroblue tetrazolium and 35 µL of 5-bromo-4-chloro-3-indolyl phosphate per 10 mL in buffer 2) in a sealed plastic bag or suitable box kept in the dark. No shaking is required. Color development is terminated by washing filters in TE buffer.

4. Notes

1. Reaction volumes for PCR using FTA filter punches should be adjusted to ensure complete immersion of the filter in the reactants. PCR on smaller punches can be performed using less volume whereas larger (6 mm diameter) punches may require 100–200 µL of reactant mixture. Filters will float toward the aqueous/oil interface during thermocycling, but that will not affect results ***(10)***.
2. Several commercially available sources of *Taq* DNA polymerase have been used in FTA filter PCR with equal success.
3. For best results and consistency, a master mix of reactants should be used. To determine the volume of each reactant for PCR, multiply the volume of one reaction to the number of reactions plus one. For example, if there are 10 reactions, multiply each reactant by 11. The master mix should contain dH_2O, buffer, dNTP, and primers. Enzyme and template are added separately.
4. PCR thermocycling conditions, such as annealing temperatures, can vary between instruments and therefore affect the amplification of DNA.
5. T_D calculation for PCR primers and DNA probes (rough estimate): Degrees in °C equals 4× the number of (C +G) plus 2× the number of (T+A).
6. For nonradioactive hybridization protocols, Boehringer Mannheim, now merged with Roche Biochemicals, and their web site has useful protocols and explanations. The site is *www.roche.com/diagnostics*.

References

1. *Bacteriological Analytical Manual* (1998) *Shigella.* Chapter 6, pp. 6.01–6.06. Gaithersburg: AOAC International.
2. Hartman, A. B.,Venkatesan, M., Oaks, E. V., and Buysee, J. M. (1990) Sequence and molecular characterization of a multicopy invasion plasmid antigen gene, *ipaH*, of *Shigella flexneri*. *J. Bacteriol.* **172,** 1905–1915.
3. Galan, J. E., Ginocchio, C., and Costeas, P. (1992) Molecular and functional characterization of the *Salmonella* gene *invA*: homology of InvA to members of a new protein family. *J. Bacteriol.* **174,** 4338–4349.
4. Grunstein, M. and Hogness, D. (1975) Colony hybridization: a method for the isolation of cloned DNAs that contain a specific gene. *Proc. Natl. Acad. Sci. USA* **72,** 3961–3965.

5. Hanes, D. E., Koch, W. E., Miliotis, M. D., and Lampel, K. A. (1995) DNA probe for detecting *Salmonella enteritidis* in food. *Mol. Cell. Probes* **9,** 9–18.
6. Rahn, K., De Grandins, S. A., Clarke, R. C., McEwen, S. A., Galan, J. E., Ginochhio, C., et al. (1992) Amplification of an *invA* gene sequence of *Salmonella typhimurium* by polymerase chain reaction as a specific method of detection of *Salmonella. Mol. Cell. Probes* **6,** 271–279.
7. Lampel, K. A., Keasler, S. P., and Hanes, D. E. (1996) Specific detection of *Salmonella enterica* serotype Enteritidis using the polymerase chain reaction. *Epidemiol. Infect.* **116,** 137–145.
8. Cha, R. S., Zarbl, H. Keohavong, P., and Thilly, W. G. (1992) Mismatch Amplification mutation assay (MAMA): application to the c-H-*ras* gene. *PCR Methods Applic.* **2,** 14–20.
9. Gulig, P. A., Caldwell, A. L., and Chiodo, V. A. (1992). Identification, genetic analysis and DNA sequence of a 7.8-kb virulence region of the *Salmonella typhimurium* virulence plasmid. *Mol. Microbiol.* **6,** 1395–1411.
10. Lampel, K. A., Kornegay, L. and Orlandi, P. A. (2000) Improved template preparation for polymerase chain reaction based assays for the detection of food borne bacterial pathogens. *Appl. Environ. Microbiol.* **66**, 4539–4542.

23

Polymerase Chain Reaction for Detection of *Listeria monocytogenes*

Debra K. Winters

1. Introduction

There are several species of *Listeria,* but only *Listeria monocytogenes* has been identified as the principal pathogen in humans and animals. *L. monocytogenes* is a ubiquitous Gram-positive bacterium responsible for an uncommon but potentially serious infection in humans who ingest contaminated food. The symptoms of listeriosis in humans can include meningitis, encephalitis, and sepsis. In some cases fatalities have occurred. The major risk groups are pregnant women, young children, elderly people, and the immune compromised. *L. monocytogenes* has been shown to survive in a wide range of environmental conditions including salt concentrations of up to 10%, temperatures as low as 4°C, and in a variety of food products *(1)*. According to the Centers for Disease Control (CDC), between 1988 and 1990, there was an annual incidence of 7.4 cases per 1 million population in the surveillance areas. Of these cases, 23% of the individuals died and of that number 33% were pregnant women. There are at least 16 serotypes of *L. monocytogenes*, which are identified by a serological grouping of the five heat-labile flagellar antigens and the 14 carbohydrate-containing heat-stable antigens *(2)*. *Listeria* are ubiquitous in nature, being found in soil, sewage, river water, vegetable matter, silage, other animal fodder, insects, the human intestines, raw milk, and kitchen premises *(3)*.

Traditional analysis of food for the presence of microorganisms relies on the growth of bacteria in artificial media. Viable cells that cannot be cultured will therefore not be recognized. In addition, culture techniques are often time consuming and not reliable *(4)*. Faster, more specific methods for detection of

From: *Methods in Molecular Biology, vol. 176: Steroid Receptor Methods: Protocols and Assays*
Edited by: B. A. Lieberman © Humana Press Inc., Totowa, NJ

Table 1
PCR Primers and Parameters for Detecting *Listeria monocytogenes*

Target gene	Primer sequences (5'-3')	Annealing temp.	Cycles	Product length	Reference
hlyA	Forward GCATCTGCATTCAATAAAGA Reverse TGTCACTGCATCTCCGTGGT	60	35	174	***18***
hylA	Forward CCTAAGACGCCAATCGAA Reverse AAGCGCTTGCAACTGCTC	50	30	702	***19***
Dth18	Forward CCGGGAGCTGCTAAAGCGGT Reverse GCCAAACCACCGAAAAGACC	54	30	326	***20***
α-hemolysin	Forward CGGAGGTTCCGCCAAAGATG Reverse CCTCCAGAGTGATCGATGTT	55	30	234	***21***
β-hemolysin	Forward ACAAGCTGCACCTGTTGCAG Reverse TGACAGCGTGTGTAGTAGCA	55	30	130	***21***
hylA	Forward AACCTATCCAGGTGCTC Reverse CGCCACACTTGAGATAT	60	35	520	***22***
pepC	Forward GGTCGGTGCATTAATAAG Reverse CAAGAGTTACAAATTACACC	52	40	90	***23***
16S rRNA	Forward CACGTGCTACAATGGATAG Reverse AGAATAGTTTTATGGGATTAG	48	40	70	***8***
hylA	Forward CATCGACGGCAACCTCGGAGA Reverse ATACAATTACCGTTCTCCACCATTC	62	30	417	***24***

Table 1. (*cont.*)

Target gene	Primer sequences (5'-3')	Annealing temp.	Cycles	Product length	Reference
hylA	Forward ATTGCGAAATTTGGTACAGC Reverse ACTTGAGATATATGCAGGAG	55	30	234	***25***
iap	Forward CGAATCTAACGGCTGGCACA Reverse GCCCAAATAGTGTCACCGCT	50	30	287	***26***

foodborne pathogens include detection with antibodies and DNA probes. In recent years methods of detection based on PCR have been developed for several pathogenic microorganisms including *L. monocytogenes* ***(5–8)***. These probes are based on target genes such as *hly*, 16S rRNA, and others. The *hly* gene codes for listerolysin O, the hemolysin activity of *L. monocytogenes* that causes the vesicles that form on infection to rupture. For this reason, it is sometimes referred to as LLO. It appears to be a major virulence factor of *L. monocytogenes* and thus a major target for polymerase chain reaction (PCR) identification to distinguish *L. monocytogenes* from other species of *Listeria*. Several excellent reviews on the pathogenicity of *Listeria monocytogenes* are available for more in-depth discussion ***(9,10)***. There are also several review articles on the PCR of foodborne pathogens and of *L. monocytogenes* ***(4–6,11)***.

2. Materials

2.1. Selecting the Primers

Primers are usually 20–25 nucleotides long with a G + C content varying from 40% to 60%. The annealing temperature can be chosen between 30 and 70°C, allowing an optimal adaptation of cycle parameters to appropriate annealing temperatures of the primers. Typically they have a melting temperature between 55°C and 80°C. **Table 1** shows various PCR primers for *L. monocytogenes* that appear in the literature. The annealing temperature for this group ranges from 48°C to 62°C. The primer must be dephosphorylated on the 3' end ***(12)***. Primers should be stored as a 10× stock in aliquots at –20°C. Optimal concentrations are 0.1–1 µM. An increase in the oligodeoxynucleotide concentration may lead to an increase in nonspecific bands such as primer duplex formation and polymerization that leads to a product of less than the length of the two primers.

2.2. Thermostable DNA Polymerase

DNA polymerase catalyzes the DNA-dependent polymerization of dNTPs. One unit of the enzyme is defined as the amount of enzyme that will incorporate 10 nmol of radioactively labeled dTTP into acid-insoluble material at 80°C in 30 min ***(13)***. The enzyme is often purified to a specific activity of 20,000 U/mg of protein so that 1 µL containing 0.5–1.25 U is enough enzyme for a 100-µL reaction. If the enzyme concentration is too low, insufficient product will be produced. If the enzyme concentration is too high, nonspecific background products may be produced. The optimal temperature is between 72 and 75°C, which minimizes secondary structure of the template, resulting in higher polymerization yield. Incorporation rates decrease above 90°C. Originally PCR was designed without the aid of a thermostable polymerase. New enzyme had to be added manually at the beginning of every cycle. A variety of thermostable DNA polymerases are now commercially available that differ in temperature stability, fidelity, and optimal reaction conditions. Thermostable DNA polymerases such as *Pfu* that possess 3'–5' exonuclease activity produce blunt-ended PCR products whereas other polymerases, such as *Taq*, which lack this proofreading activity generate PCR products containing a single 3' adenosine nucleotide extension ***(14)***.

There has been a long-lasting lawsuit over the licensing of *Taq* polymerase. For this reason, companies must purchase a license to sell this product. Thermostable polymerases can be purchased from the following companies among others: Amersham Pharmacia Biotech, Boehringer Mannheim, CLONTECH Laboratories, DNAmp (U.K.), Enzyme Technologies, Epicentre Technologies, Fermentas AB, Finnzymes OY, Fisher Scientific, New England Biolabs, Life Technologies, PE Applied Biosystems, Promega, QIAGEN, Sigma-Aldrich, Stratagene, and TaKaRa Shuzo.

2.3. Buffers

The buffer composition used in PCR can vary. Many companies supply the buffer that they recommend with the enzyme. Buffers can also be made by adding individual components. A common recipe would be 10 m*M* Tris-HCl, pH 8.8, 50 m*M* KCl; and 1.5–2.5 m*M* $MgCl_2$. The concentration of the Tris or phosphate buffer is not crucial but should be kept low. The concentration of the $MgCl_2$ can be critical and is discussed in more detail in a later section.

2.4. Deoxynucleotides

Maximal polymerization rates are obtained with approx 0.25 m*M* dNTPs. Substrate inhibition is observed at dNTP concentrations of 4–6 m*M*. The deoxynucleotides at a concentration 100 m*M* are usually provided by the companies that sell the thermostable polymerase. The four deoxynucleotides are mixed at an equal ratio so that the concentration of each in the mix is 2.5 m*M*. The dNTPs should be used at equimolar levels to minimize misincorporation. As dNTPs can chelate free magnesium ions, excess dNTPs should not be added to PCR unless compensated by adding additional Mg^{2+}.

2.5. $MgCl_2$

The enzyme activity of the DNA polymerase is dependent on bivalent cations. Concentrations of 2 m*M* $MgCl_2$ are optimal for most systems but some systems have been shown to be better at higher concentrations. The concentration can be dependent on the specific thermostable DNA polymerase used. The reaction should contain 0.5–2.5 m*M* above that of the total dNTP concentration ***(15)***. However, as the Mg^{2+} concentration increases, the amplification of unspecific sequences also increases. Monovalent cations also have an effect on the activity of the enzyme. Optimum conditions are 50 m*M* KCl, whereas inhibition is reported at concentrations above 75 m*M* KCl ***(16)***.

2.6. Template

The sensitivity of the method allows for a theoretical limit of one molecule per test. Normally less than 1 μg of DNA is used for an experiment, and, depending on the source, as little as 10 pg–100 ng are required. DNA template from a variety of sources ranging from intact cells to purified DNA can be used. The cells are lysed either prior to placing in the thermocycler or in the first step of denaturation. The presence of a nonionic detergent such as Triton X-100 is to make the DNA more accessible (*see* **Note 1**).

2.7. Thermocycler

PCR requires repeated cycling through a series of temperatures to denature, anneal, and elongate. Sample volumes up to 100 μL are used in small reaction tubes. Tubes can be specially purchased for PCR that have a thinner wall and thus allow better heat transfer from the incubation device. If these tubes are not available, ordinary 500-μL microcentrifuge tubes may be used (*see* **Note 2**).

2.8. Components that Make Up the Gel

2.8.1. Agarose TBE/TAE Buffer, Ethidium Bromide, Loading Buffer

Any molecular biology agarose is appropriate. Companies such as FMC Bioproducts (Rockland, ME) carry a variety of types of agarose that are appropriate for different size PCR products. The concentration of agarose used in the gel is dependent on the size product expected. For products such as those listed in **Table 1**, ranging from 90 bp to 700 bp, a 1.5%–1% gel would be run. Ethidium bromide is prepared as a 10 mg/mL stock. To a 100-mL gel, 1.5 µL of ethidium bromide are added. Recipes for TAE and TBE buffer are as follows *(17)*:

1. 40X TAE buffer: Combine 193.6 g of Tris base, 108.9 g of sodium acetate·3 H_2O, 15.2 g $Na_2EDTA·2H_2O$, and 700 mL of water. Dissolve, and adjust the pH to 7.2 with acetic acid. Add water to a final volume of 1 L and autoclave. Store at room temperature.
2. 20X TBE: Combine 121 g of Tris base, 61.7 g of boric acid, 7.44 g $Na_2EDTA·2H_2O$, and water to a final volume of 1 L. A precipitate invariably forms in this solution at room temperature. The buffer continues to work well despite the precipitate.
3. 5X Gel-loading buffer: Combine 5 mL of glycerol, 1 mL 10× TBE, 1 mL of 10% bromophenol blue, 1 mL of 10% xylene cyanol, and 2 mL of water. Mix and store at 4°C in aliquots.

3. Methods

3.1. Sterilize All Equipment

All tubes, pipet tips, pipets, etc. should be sterilized before beginning. Gloves should be worn throughout the entire procedure. Sterile technique is observed with all steps. Some laboratories that have the space set up an entire room devoted solely to PCR. Many laboratories have glassware dedicated to PCR use only or have purchased a hood with a UV light source that sterilizes reagents prior to use. Pipet tips with a barrier aerosol tip can be purchased and used for PCR studies to ensure no cross-contamination from the pipets. When thawing individual components, place them on ice and allow to thaw slowly. Keep all solutions on ice during the setup of the PCR assay.

3.2. Preparation of Template

Listeria monocytogenes cells are grown to late log stage in a liquid broth such as brain heart infusion (BHI). This usually takes 16–24 h. *L. monocytogenes* does not have fastidious culture requirements and will grow rapidly on many commonly used media. The cells are then centrifuged (1-mL aliquot) in a microcentrifuge at 12,000 rpm for 5 min. The supernatant is discarded and the pellet is resuspended in 100 µL of 1% Triton X-100. This is boiled for 5 min to lyse the cells. This lysing step can be omitted and the cells used directly, as they will lyse in the first denaturation of the thermocycler step. Otherwise, centrifuge briefly to ensure a clear supernatant layer (*see* **Notes 3–5**).

3.3. PCR Setup

A typical PCR assay contains the following per 25-µL reaction:

Component	µL
10× buffer (supplied by the manufacturer)	2.5
$MgCl_2$ (25 m*M*) (supplied)	2.5
dNTPS	2.0
Primer 1	1.5
Primer 2	1.5
Template	1.5
Taq polymerase	0.5
Sterile water	13

It is easiest to make a master mix of all components for the number of tubes to run and to dispense this into the number of tubes you need rather than to pipet the above amounts into each separate tube. Some companies, such as Gibco BRL Life Technologies, now sell a premix of all components except the primers and the DNA template. Use of such a premix simplifies the reaction and allows for less chance of leaving a single component out (*see* **Note 6**).

3.4. Thermocycler

Once the reaction components are mixed and dispensed into the appropriate number of tubes, centrifuge each briefly to ensure all of the mix is at the bottom of the tube. Then add a drop of mineral oil into the middle of the tube. Place the tube into the thermocycler and run the program as dictated by the chosen primer (**Table 1**). A typical program looks like this:

Denature at 90–95°C for 1–7 min

20–40 Cycles of: **Denature** 10 s to 1 min at 92–95°C

Anneal 10 s to 1 min at temperature determined by primer specificity but should be within a range of 5 degrees above or below the T_m for the primers.

Elongate 30 s–1 min at 72–75°C depending on the length of the expected PCR product and the particular thermostable enzyme used.

This is typically followed by a 5–10-min final extension step, which is often beneficial to increase the yield of the full-length products. Most reactions take approx 3 h depending on the instrument used and how fast it changes from one temperature to the next. This time can be calculated and you can return at the end of the cycle. When PCR was first described, the thermocycler that changes temperatures quickly and automatically did not exist. A person had to physically sit in the laboratory and manually move the tubes from one water bath set at the first temperature to the next water bath set at the second temperature. This had to be done for the entire 30–40 cycles.

3.5. Preparation of Samples for Gel Electrophoresis

Remove 10 μL of the sample in the lower layer beneath the oil. Place this in a 0.5-mL microcentrifuge tube. Add 2.0 μL of loading buffer to the sample. Centrifuge to ensure all contents are at the bottom of the tube. Load the samples along with molecular weight markers in the wells of a gel of appropriate agarose, depending on the size of the products made. Markers of varying size can be purchased from any one of the companies that sells molecular biology reagents. For small products, such as the 90 basepair (bp) product from the aminopeptidase primers (**Table 1**), the Amplisize markers from Bio-Rad work well.

3.6. Running the Gel

Run the gel that either contains ethidium bromide or stain with ethidium bromide after running the gel. Visualize with a transilluminator. The gel can be prepared by dissolving agarose in either 1× TAE or 1× TBE buffer using a microwave. If a microwave is not available, put the agarose and buffer in a 250-mL Erlenmeyer flask. Take a 25-mL Erlenmeyer flask and turn it upside down inside the 250-mL flask. Heat this to boiling while stirring. Allow the agarose to cool to 60°C before adding the ethidium bromide. Pour and allow to solidify. The samples are loaded as with any submarine horizontal agarose gel and run. Visualize with the transilluminator and photograph with a Polaroid Quick Shooter Camera (*see* **Notes 7** and **8**).

4. Notes

1. Sensitivity. The reported sensitivity of the technique has varied from author to author. We find that the sensitivity is usually 100–1000 cells. Other authors have reported a wide variety ranging from 1 cell to 10,000. A number of factors including food sample preparation could be contributing to the wide range reported. The number of *L. monocytogenes* required to initiate an infection in an individual is not known. For this reason it is hard to say what assay sensitivity is required to make the PCR useable in the real world.
2. Thermocyclers. There are many thermocyclers on the market these days. We use a dual block from Ericomp. The advantage of the dual block is that more than one program can be run at a time. There are also different block formats, including 0.5-mL tubes, 0.2-mL tubes, and 96-well plates. Another consideration when ordering a thermocycler is the option of a heated lid. This recent addition to thermocyclers does away with the need to add oil to your sample. As the lid is heated, the sample will not evaporate.

3. Time scale. The PCR itself can be done in one working day. It takes <2 to harvest the cells, lyse them, and prepare the PCR. It then takes up to 4 h to run the PCR depending on how many cycles are in the program. Another hour or two is devoted to pouring and running the gel. The total is less than 8 h. The cells must be grown before starting, which perhaps takes up to 48 h. Depending on the type of bacteria, enrichment steps may also make the procedure longer.
4. Variations to the basic system. Several things can be substituted from the system described here. We use Triton X-100 as the detergent but 0.5% SDS followed by a digestion with pronase K can also be employed ***(28)***. The initial boiling step may be omitted, as the cells will lyse during the first heating of 95°C. Described here is the growth of *Listeria* in liquid medium but it is possible to perform PCR from a single colony. Pick a colony and place it directly into the PCR tube containing the rest of the mix. It will lyse when the first heating step occurs.
5. Problems with food samples. PCR is developed with an organism or group of organisms in mind. Therefore, we have concentrated on the "how-to" of pure culture. However, in the real world, food pathogens, such as *Listeria* will be found contaminating a food specimen. For *L. monocytogenes*, milk and dairy products were the first among the studied foods. This food pathogen has been found in 2 to 5% of raw milk samples ***(29)***. A wide variety of cheeses have been found to be contaminated as have meats, meat products, different types of vegetables and seafood ***(30–32)***.

   There are many factors in food samples that may provide interference with the PCR assay you have chosen. Interesting strategies to overcome food interference problems have been reported including washing the target cells followed by centrifugation prior to the PCR assay ***(33)***, filter membranes ***(34)***, the use of antibodies, latex bead antibody or magnetic immunobead antibody capture of target cells ***(35,36)***, and antibody to the RNA–DNA complex ***(37)***.

   One question that frequently comes up in regard to the utility of PCR in food samples is that of viability of the cells and their potency as pathogens. If the cells are not alive PCR will still detect them but perhaps they no longer pose a threat. It has been suggested that one potential method to circumvent this is to use RNA in the initial reaction. As RNA has a shorter half-life than DNA, this might be a better indicator of viability of pathogenic cells, which might be present in the food sample. One could easily, in the presence of reverse transcriptase, convert the RNA into DNA before the amplification begins.

   Frequently a problem encountered when working with food samples is isolation of the organisms of interest. Foodstuffs contaminated with pathogens such as *Listeria* can be homogenized in a stomacher, centrifuged at a low speed to remove the food debris, and then centrifuged at a higher speed to pellet the bacteria. Other methods of preparation of the food sample include magnetic bead DNA extraction ***(38)*** and centrifugation followed by heat treatment ***(8)***. If one does need to isolate the organism, it will be necessary to employ selective enrichment broths and selective differential agars ***(39)***. An alternative approach is to "cold enrich" samples by incubating them at 4°C for an extended period. This technique takes advantage of the fact that *Listeria* can multiply at low temperatures, whereas most other contaminating organisms cannot. However, growth at this temperature is very slow so that this technique can take several weeks to months ***(40).***

6. Primer selection. There are many primers in the literature for *L. monocytogenes.* Some of the primer sets are for the same gene but target a different area of the gene, such as *hyl*A. A few are listed in **Table 1**. The listeriolysin O is also a popular target as is the 16S rRNA sequence. In many cases, the primers were tested with only a few species of *Listeria.* However, one set of primers based on a listeriolysin O sequence proved to detect only *Listeria* species in 95 out of 95 cases ***(27)***. The sequence of the primers, however, was not revealed. Other genes have been targeted such as an *aminopeptidase C* and the *Dth18* gene ***(20,23)***. A set of primers based on *Dth18* gene fragment gave amplification of 326 product and detection in culture lysates with a sensitivity of approx 1–10 colony-forming units (CFU) ***(20)***. In food samples 1000 CFU were required for detection. Primer sets based on the α- and β-hemolysin genes were used in pure culture with sensitivities of 10 bacteria for the α-hemolysin and 50 bacteria for the β-hemolysin ***(21).***
7. What to do if you do not see bands. If you fail to obtain bands from the PCR then first check all your reagents and make sure that they are not too old. Fresh preparations or fresh dilutions of concentrated stocks sometimes overcome problems encountered. Sometimes there is insufficient PCR amplification owing to a shortage of *Taq* polymerase, nucleotides, or magnesium. Frequently inhibitors in the foods block the annealing or inhibit the amplification. Purification of the template DNA may be necessary. The amplification of difficult templates can sometimes be overcome by including additional reagents such as dimethyl sulfoxide (DMSO), betaine, formamide, or single-stranded DNA binding protein ***(41).*** Other methods that can be attempted are the use of chaotropic salts and inorganic solvents to extract the DNA and obtain a more purified preparation.
8. What if you get too many bands? A method of preventing nonspecific priming employs an initial annealing temperature above that of the primers, and the temperature is lowered in each cycle until a product is generated. This is referred to as "touchdown" PCR ***(42).*** Preamplification heating may also increase the specificity of the reaction ***(43).*** Holding the reaction at a higher temperature may reduce primer dimer formation as well as foster more stringent primer annealing during the first round of elongation. As mentioned in an earlier section, the magnesium concentration can be critical and should be optimized for each set of primers to minimize nonspecific priming. In recent years, several companies have begun to develop thermostable enzymes bound to an antibody, such as platinum *Taq* DNA polymerase by Gibco BRL Life Technologies (Rockville, MD). This is a recent modification of the popular idea of "hot start" PCR ***(44).*** The idea is to withhold one of the essential components of the system until all the other components have been heated above the annealing temperature. The missing component is then added by manual pipetting or by melting a wax barrier that physically

separates the essential components within the reaction tubes. Another recent modification is that companies supply all reagents in the form of a "bead" or liquid mix. The researcher adds their specific template and primers. This helps to reduce the amount of variability of the reaction and can produce more reproducible bands. Very recently an article appeared describing a method of Hot-Start PCR the addition of double-stranded DNA fragments are adapted to the annealing temperature and help to prevent nonspecific binding ***(45)***. Commercially available "kits" for PCR have increased dramatically over the past 5 yr. Epicentre Technologies sells a "FailSafe PCR PreMix Selection Kit." This would be a good place to start when trying to optimize conditions for a newly developed PCR method. Otherwise, a set of primers may be chose from **Table 1** and PCR for *L. monocytogenes* begun under the described conditions.

References

1. Amezaga, M-R., Davidson, I., McLaggan, D., Verheul, A., Abee, T., and Booth, I. R. (1995) The role of peptide metabolism in the growth of *Listeria monocytogenes* ATCC 23074 at high osmolarity. *Microbiology* **141,** 41–49.
2. Seeliger, H. P. R. and Hohne, K. (1979) Serotyping of *Listeria monocytogenes* and related species. In *Methods in Microbiology*, vol. 13 (Bergan, R. and Norris, J. R., eds.), Academic Press, Orlando, FL, pp. 31–49.
3. Lacey, R. W. (1993) Food-borne bacterial infections. *Parasitology* **107,** S75–S93.
4. Johnson, M. G., Wang, R-F., and Winters, D. K. (1997) PCR and nucleic acids amplification methods. In *Food Microbiological Analysis: New Technologies* (Tortorello, M. L. and Gendel, S. M., eds.), Marcel Dekker, New York, pp. 183–205.
5. Allmann, M., Hofelein, C., Koppel, E., Luthy, J., Meyer, R., Niedenhauser, C., et al. (1995) Polymerase chain reaction (PCR) for detection of pathogenic microorganisms in bacteriological monitoring of dairy products. *Res. Microbiol.* **146,** 85–97.
6. Candrian, U. (1995) Polymerase chain reaction in food microbiology. *J. Microbiol. Meth.* **23,** 89–103.
7. Simon, M. C., Gray, D. I., and Cook, N. (1996) DNA extraction and PCR methods for the detection of *Listeria monocytogenes* in cold-smoked salmon. *Appl. Environ. Microbiol.* **62,** 822–824.
8. Wang, R-F., Cao, W-W., and Johnson, M. G. (1992) 16S rRNA-based probes and polymerase chain reaction method to detect *Listeria monocytogenes* cells added to foods. *Appl. Environ. Microbiol.* **58,** 2827–2831.
9. Farber, J. M. and Peterkin, P. I. (1991) *Listeria monocytogenes*, a food-borne pathogen. *Microbiol. Rev.* **55,** 476–511.
10. Portnoy, D. A., Chakroaborty, T., Goebel, W., and Cossart, P. (1992) Molecular determinants of *Listeria monocytogenes* pathogenesis. *Infect. Immun.* **60,** 1263–1267.
11. Hill, W. E. (1996) The polymerase chain reaction: applications for the detection of foodborne pathogens. *Crit. Rev. Food Sci. Nutr.* **36,** 123–173.

12. Gelfand, D. H. (1989) *Thermus aquaticus* DNA polymerase. In *Current Communications in Molecular Biology: Polymerase Chain Reaction* (Erlilch, H. A., Gibbs, R., and Kazazian, H. H., eds.), Cold Spring Harbor Laboratory, Cold Spring Harbor, NY, pp. 11–17.
13. Chien, A., Edgar, D. B., and Trela, J. M. (1976) Deoxyribonucleic acid polymerase from the extreme thermophile *Thermus aquaticus. J. Bacteriol.* **127,** 1550–1557.
14. Clark, J. M. (1988) Novel non-templated nucleotide addition reactions catalyzed by prokaryotic and eucaryotic DNA polymerases. *Nucleic Acids Res.* **16,** 9677–9686.
15. Innis, M. A. Gelfand, D. H., Sninsky, J. J., and White, T. J. (1990) *PCR Protocols: A Guide to Methods and Applications.* Academic Press, San Diego, CA.
16. Innis, M. A., Myambo, K. B., Geifand, D. H., and Brow, M. A. D. (1988) DNA sequencing with *Thermus aquaticus* DNA polymerase and direct sequencing of polymerase chain reaction-amplified DNA. *Proc. Natl. Acad. Sci. USA* **85,** 9436–9440.
17. Davis, L., Kuehl, M., and Battey, J. (1994) *Basic Methods in Molecular Biology.* Appleton & Lange, Norwalk, CT.
18. Deneer H. G. and Boychuk, I. (1991) Species-specific detection of *Listeria monocytogenes* by DNA amplification. *Appl. Environ. Microbiol.* **57,** 606–609.
19. Border, P. M., Howard, J. J., Plastow, G. S., and Siggens, K. W. (1990). Detection of *Listeria* species and *Listeria monocytogenes* using polymerase chain reaction. *Lett. Appl. Microbiol.* **11,** 158–162.
20. Wernars, K., Heuvelman, C. J., Chakraborty, T., and Nortermans, S. H. W. (1991) Use of the polymerase chain reaction for direct detection of *Listeria monocytogenes* in soft cheese. *J. Appl. Bacteriol.* **70,**121–126.
21. Furrer, B., Candrian, U., Hoefelein, C., and Luethy, J. (1991) Detection and identification of *Listeria monocytogenes* in cooked sausage products and in milk by *in vitro* amplification of haemolysin gene fragments. *J. Appl. Bacteriol.* **70,** 373–379.
22. Thomas, E. J. G., King, R. K., Burchak, J., and Gannon, V. P. J. (1991) Sensitive and specific detection of *Listeria monocytogenes* in milk and ground beef with the polymerase chain reaction. *Appl. Environ. Microbiol.* **57,** 2576–2580.
23. Winters, D. K., Maloney, T. P., and Johnson, M. G. (1999). Rapid detection of *Listeria monocytogenes* by a PCR assay specific for an aminopeptidase. *Mol. Cell. Probes* **13,** 127–131.
24. Fitter, S., Heuzenroeder, M., and Thomas, C. J. (1992) A combined PCR and selective enrichment method for rapid detection of *Listeria monocytogenes. J. Appl. Bacteriol.* **73,** 53–59.
25. Golsteyn Thomas, E. J., King, R. K., Burchak, J., and Gannon, V. P. J. (1991) Sensitive and specific detection of *Listeria monocytogenes* in milk and ground beef with the polymerase chain reaction. *Appl. Environ. Microbiol.* **57,** 2576–2580.

26. Jaton, K., Sahli, R., and Bille, J. (1992) Development of polymerase chain reaction assays for detection of *Listeria monocytogenes* in clinical cerebrospinal fluid samples. *J. Clin. Microbiol.* **30,** 1931–1936.
27. Bessesen, M. E., Luo, Q., Rotbart, H. A., Blaser, M. J., and Ellison, R. T. (1990) Detection of *Listeria monocytogenes* by using the polymerase chain reaction. *Appl. Environ. Microbiol.* **56,** 2930–2932.
28. Allman, M., Hofelein, C., Koppel, E., Luthy, J., Meyer, R., Niederhauser, C., et al. (1995) Polymerase chain reaction (PCR) for detection of pathogenic microorganisms in bacteriological monitoring of dairy products. *Res. Microbiol.* **146,** 85–97.
29. Farber, J. M., Sanders, G. W., and Malcolm, S. A. (1988) The presence of *Listeria* spp. in raw milk in Ontario. *Can. J. Microbiol.* **34,** 476–511.
30. Johnson, J. L., Doyle, M. P., and Cassens, R. G. (1990) Incidence of *Listeria* spp. in retail meat roasts. *J. Food Sci.* **55,** 572–574.
31. Heisick, J. E., Wagner, D. E., Nierman, M. L., and Peeler, J. T. (1989) *Listeria* spp. found on fresh market produce. *Appl. Environ. Microbiol.* **55,** 1925–1927.
32. Farber, J. M. (1991) *Listeria monocytogenes* in fish products. *J. Food Protect.* **54,** 922–934.
33. Winters, D. K., O'Leary, A. E., and Slavik, M. F. (1998) Polymerase chain reaction for rapid detection of *Campylobacter jejuni* in artificially contaminated foods. *Lett. Appl. Microbiol.* **27,**63–167.
34. Kirk, R. and Rowe, M. T. (1994) A PCR assay for detection of *Campylobacter jejuni* and *Campylobacter coli* in water. *Lett. Appl. Microbiol.* **19,** 19301–19303.
35. Fluitt, A. C., Torensma, R., Visser, M. J. C., Aaarsman, C. J. M., Poppelier, M. J. J. G., Keller, B. H. I., et al. (1993) Detection of *Listeria monocytogenes* in cheese with the magnetic immuno-polymerase chain reaction assay. *Appl. Environ. Microbiol.* **59,** 1289–1293.
36. Fluitt, A. C., Widjojoatmodjo, M. N., Box, A. T. A., Toensma, R., and Verhoef, J. (1993) Rapid detection of *Salmonella* in poultry with the magnetic immuno-polymerase chain reaction assay. *Appl. Environ. Microbiol.* **59,** 1342–1346.
37. Blais, B. W. (1994) Transcriptional enhancement of the *Listeria monocytogenes* PCR and simple immunoenzymatic assay of the product using antiRNA:DNA antibodies. *Appl. Environ. Microbiol.* **60,** 348–352.
38. Doyle, M. P. and Shoeni, J. L. (1986) Selective-enrichment procedure for isolation of *Listeria monocytogenes* from fecal and biologic specimens. *Appl. Environ. Microbiol.* **51,** 1127–1129.
39. Niederhauser, C., Luthy, J., and Candrain, U. (1994) Direct detection of *Listeria monocytogenes* using paramagnetic bead DNA extraction and enzymatic DNA amplification. *Mol. Cell Probes* **8,** 223–228.
40. Gray, M. L., Stafseth, H. J., and Thorp, F. (1948) A new technique for isolating *listerellae* from the bovine brain. *J. Bacteriol.* **55,** 873–884.
41. Oshima, R. G. (1992) Single-stranded DNA binding protein facilitates amplification of genomic sequence by PCR. *BioTechniques* **13,** 188.

42. Don, R. H., Cox, P. T., Wainwright, B. J., Baker, K., and Mattick, J. A. (1991) "Touchdown" PCR to circumvent spurious priming during gene amplification. *Nucleic Acids Res.* **19,** 4008.
43. D'Aquila, R. T., Bechtel, L. J., Videler, J. A., Eron, J. J., Gorczyca, P., and Kaplan J. C. (1991) Maximizing sensitivity and specificity of PCR by pre-amplification heating. *Nucleic Acids Res.* **19,** 3749.
44. Ou, C.-Y., Kwok, S., Mitchell, S.W., Mack, D.H., Sninsky, J.J., Krebs, J.W., et al. (1988) DNA amplification for direct detection of HIV-1 in DNA of peripheral blood mononuclear cells. *Science* **238**, 295-297.
45. Kainz, P., Schmiedlechner, A., and BerndStrack, H. (2000) Specificity-enhanced Hot-Start PCR: addition of double-stranded DNA fragments adapted to the annealing temperature. *BioTechniques* **28,** 278–282.

Index

H

I

M

N

MIX
Papier aus verantwortungsvollen Quellen
Paper from responsible sources
FSC® C105338

If you have any concerns about our products,
you can contact us on
ProductSafety@springernature.com

In case Publisher is established outside the EU,
the EU authorized representative is:
Springer Nature Customer Service Center GmbH
Europaplatz 3, 69115 Heidelberg, Germany

Printed by Libri Plureos GmbH
in Hamburg, Germany